More Urban Water: Design and Management of Dutch Water Cities

Urban Water Series

ISSN 1749-0790

Series Editor:

Čedo Maksimović
Department of Civil and Environmental Engineering
Imperial College
London, United Kingdom

Volume 10

More Urban Water: Design and Management of Dutch Water Cities

Edited by

Fransje Hooimeijer
Department of Urban Design
Faculty of Architecture
Delft University of Technology
Delft, The Netherlands

Wout van der Toorn Vrijthoff
Department of Real Estate and Project Management
Faculty of Architecture
Delft University of Technology
Delft, The Netherlands

Taylor & Francis
Taylor & Francis Group

LONDON / LEIDEN / NEW YORK / PHILADELPHIA / SINGAPORE

Colophon

Book Series Editor:
Čedo Maksimović

Volume Editors:
Fransje Hooimeijer and Wout van der Toorn Vrijthoff

Editor (Dutch) and picture editing:
MaartenJan Hoekstra

Assistants:
Erwin Heurkens and Elena Mugica De La Morena

Translation:
Taalcentrum-VU, Amsterdam

Financial support:
Faculty of Architecture, Delft University of Technology, Project *Verrijking door samenwerking*
Water Research Centre, Delft University of Technology
Department of Real Estate and Housing, Faculty of Architecture, Delft University of Technology

Cover illustration

Rotterdam, Fransje Hooimeijer

Taylor & Francis is an imprint of the Taylor & Francis Group, an informa business

Typeset by Charon Tec Ltd (A Macmillan company), Chennai, India
Printed and bound in Hungary by Uniprint International (a member of the Giethoorn Media-group), Székesfehévár.

Published by: Taylor & Francis/Balkema
P.O. Box 447, 2300 AK Leiden, The Netherlands
e-mail: Pub.NL@tandf.co.uk
www.balkema.nl, www.taylorandfrancis.co.uk,
www.crcpress.com

British Library Cataloguing in Publication Data
A catalogue record for this book is available from the British Library

Library of Congress Cataloging-in-Publication Data

More urban water : design and management of Dutch water cities / Edited by Fransje Hooimeijer, Wout Toorn van der Vrijthoof.
p. cm. — (Urban water series ; ISSN 1749-0790 v. 10)
Includes bibliographical references and index.
ISBN 978-0-415-45358-5 (hardcover : alk. paper) — ISBN 978-0-203-93850-8 (e-book)
1. Municipal water supply—Management—Netherlands. I. Hooimeijer, Fransje. II. Toorn Vrijthoff, W. van der.

TD201.M67 2007
628'.21—dc22

2007032430

ISBN 978-0-415-45358-5 (hardback)
ISBN 978-0-203-93850-8 (e-book)
Urban Water Series: ISSN 1749-0790
Volume 10

Contents

List of Figures

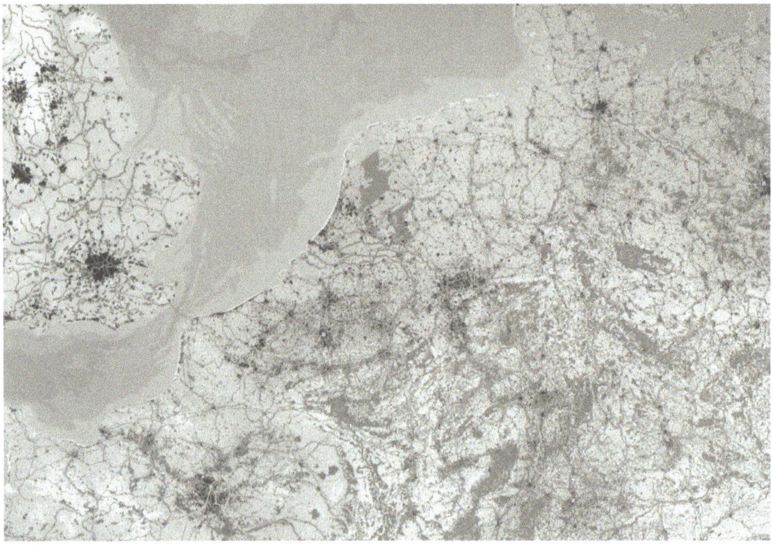

Figure 0.1 The Rhine delta.

Source: Vijfde Nota over de Ruimtelijke Ordening 2000.

Figure 0.2 The Netherlands and the cities covered by the case studies.

Source: Vijfde Nota over de Ruimtelijke Ordening 2000.

Chapter 1

Introduction: water's changing context

Wout van der TOORN VRIJTHOFF,[1] Frans van de VEN[2]

[1] *Department of Real Estate and Project Management, Faculty of Architecture, Delft University of Technology, Delft, The Netherlands*

[2] *Department of Water Management, Faculty of Civil Engineering, Delft University of Technology, Delft, The Netherlands & Division of National Water Management, Institute for Inland Water Management and Waste Water Treatment*

1.1 INTRODUCTION

About half of the Netherlands lies below sea level. This makes it a less than attractive place for human settlement. The Romans had good reason to call their province 'Germania Inferior'. Nonetheless, people did settle in what was later to become 'the Netherlands'. The original settlements were built on natural high ground, but water was still a vital factor in their situation and structure. Water was an important mode of transport for

Figure 1.1 The Amsterdam 'ring of canals'.

Source: Fransje Hooimeijer.

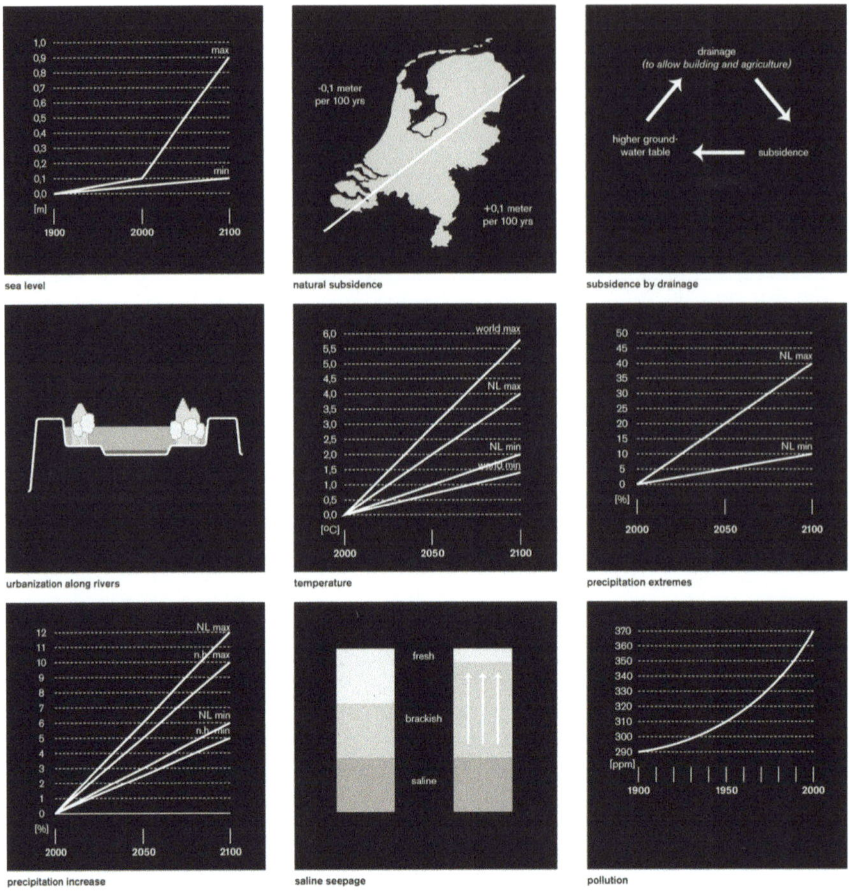

Figure 1.2 The nine problems with water in the Netherlands.

Source: Hooimeijer *et al.* 2005.

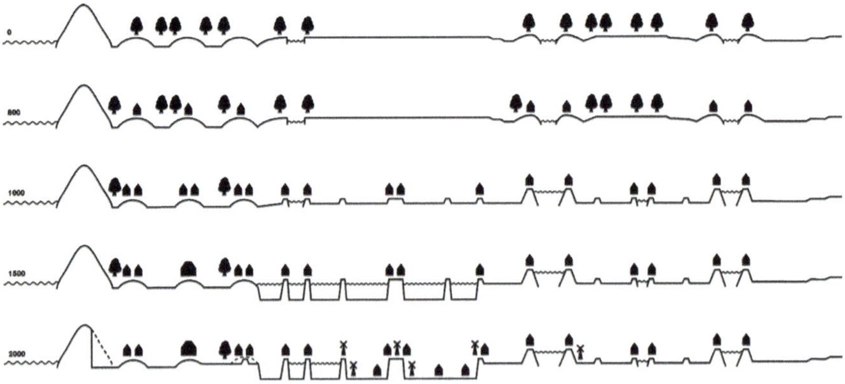

Figure 1.3 A series of sections of the Netherlands showing the natural subsidence.

Source: Hooimeijer *et al.* 2005.

goods. Moreover, watercourses were used as the drinking water supply, and wastewater was discharged into the same watercourses.

The Dutch applied technical innovations with ingenuity and started to settle into the lower-lying areas too. Pressure to do so came from the nineteenth century economic boost, as the transition from agriculture and trade to an industrial society coincided with population growth.

Traces of the structure originally created to adapt areas outside the city to agriculture are still visible in the design of the nineteenth century urban expansions. For reasons of hygiene, another innovation in this period was the separation of the drinking water supply, the removal of wastewater, and the urban water systems for keeping the ground dry underfoot. The separation of the water chain (drinking water, wastewater and treatment) from the water system (rainwater and surface water) progressed constantly.

The ensuing period, after World War II, saw unprecedented growth of the urban area. Indeed, some 75% of the current urban area was built in the post-war period. New techniques for clearing sites allowed urban developers to escape from the agriculture-based zoning structure and the geographical properties of the land.

The Dutch have an advanced urban water system, which has developed over the centuries to suit a climate that has remained fairly stable throughout that time. Recently, however, climate changes have become apparent, which are putting pressure on the urban water system. Furthermore, the sustained economic growth was accompanied by rising standards in living. Constant urbanisation itself increases the risk of flooding, which is all the more relevant as society becomes less willing to accept risks.

The greater demands now being placed on the water system and the living environment necessitate adjustments in the spatial structure of the urban area. However, different kinds of adjustments are called for in the various parts of the city. For instance, the historic city centre has entirely different spatial characteristics from the urban areas built in the period between 1850 and 1940, and both differ entirely from the post-war neighbourhoods built after 1945.

Water is not the only determinant of the spatial functional adjustments: another is the market position of the urban areas. The post-war residential districts no longer satisfy the current demand on the housing market, which means that these neighbourhoods need to be *restructured.* Sometimes the restructuring involves demolition and new building, and sometimes alteration and improvement of the still viable homes and public spaces. The restructuring costs, including the approach to the urban water issue, are normally incorporated into a new financial operation.

The market position of neighbourhoods built between 1850 and 1940 is better than that of their post-war counterparts. There is no need for drastic repositioning, nor for changes in the property and urban structure. Although the market considerations militate for *consolidation*, something still has to be done about the water issue.

The position of the historic city centre has changed substantially in recent decades. It was once the place where all urban services were concentrated. Scale expansion and accessibility requirements have caused a large-scale reorganisation of businesses and institutions that were originally based in the old centre. While these developments were taking place, public awareness of cultural heritage was growing rapidly. The cultural dimension and its significance for the identity of the city became dominant in the profile of the historic city centre. The historically anchored identity resides in the buildings and the urban structure, and the significance of water has traditionally been crucial. Making the historic city centre the actual showcase of the city demands a *transformation* of the area: an approach based on small-scale interventions compatible with the differentiated ownership structure of the property and the fine-grained nature of the urban structure.

This book looks at the integration of the solution to the urban water problem with the upgrading of property and urban structures in the existing city. From the perspective of the market, the upgrading can be classified according to the nature and scale of the intervention concerned, linked to the types of areas. The objective is to suggest guidelines for how to integrate the solution to the urban water problem into plans for the recovery (transformation), preservation (consolidation) and renewal (restructuring) of existing urban areas. The focus is on spatial, functional and financial feasibility aspects.

This chapter explains the urban water problem. Chapters 2, 3 and 4 place the problem in a broader context, addressing the characteristics of urban water, the urban development issues in the existing city and the specific water issues. Chapters 5, 6 and 7 discuss case studies within a framework formed by the trinity of transformation, consolidation and restructuring.

1.2 CLIMATE CHANGE

A clear but gloomy scenario is being projected by some observers: as the earth heats up, sea levels will rise considerably. As this happens, the polar areas will absorb more heat, weakening the warm Gulf Stream and making Europe distinctly colder.

The North Pole ice will melt first. This will actually leave the sea level unchanged, because ice occupies more volume than water. However, the reduced salinity of the water could alter the currents. The recently predicted rise in sea level, approximately 0.60 metres in 2070, is mostly attributable to the expansion of the water in the seas and oceans. Any spectacular rises will come later when the land ice on Antarctica and Greenland melt or crumble away, adding to the volume of water. The amount of land ice on Antarctica and Greenland is so large that if it were all to end up in the oceans, the sea level would rise, in theory, by 69 metres! This possibility is not yet reflected in the scenarios in this book. The Atlantis scenario involves part of the land ice breaking off as expansion of the seawater increases the pressure on the ice.

However, the question that remains is whether the climate is undergoing a brief fluctuation, or is somewhere in the middle of a development that started long ago. In search of the answer, Terje Loyning of the Norwegian Polar Institute in Tromsø has studied old logbooks, ships' logs and marine charts, going back as far as 1553. These have led him to conclude that the North Pole ice started to recede before the industrial revolution. It is a natural variation, which might also have something to do with human activities.

The United Nations Intergovernmental Panel on Climate Change (IPCC) has also carried out research into temperature fluctuations. They conclude that '*the global average surface temperature (the average of near surface air temperature over land, and sea surface temperature) has increased since 1861. Over the 20th century the increase has been 0.6 ± 0.2 °C.*' (IPCC 2001).

Although the researchers set out to clarify the causes and backgrounds, the general population was confronted with facts. The facts pointed to a clear conclusion: the climate is changing conspicuously. The changes are not restricted to specific regions but can be observed worldwide, although the effects differ from one region to another.

2050		G	G+	W	W+
Changing circulation West Europe		No	Yes	Yes	No
Winter	Average precipitation	+4%	+7%	+7%	+14%
Summer	Average precipitation	+3%	-10%	+6%	-19%
Summer	Evaporation	+3%	+8%	+7%	+15%
Sea level	Absolute raise	15-25 cm	15-25 cm	20-35 cm	20-35 cm

Figure 1.4 Four climate scenarios for the Netherlands. G = Moderate 1°C temperature rise on earth in 2050 compared to 1990, no change in air circulation patterns in Western Europe. G+ = Moderate + 1°C temperature rise on earth in 2050 compared with 1990, + milder and wetter winters because of more westerly winds, + warmer and drier summers because of more easterly winds. W = Warm 2°C temperature rise on earth in 2050 compared with 1990, no change in air circulation patterns in Western Europe. W + = Warm + 2°C temperature rise on earth in 2050 compared with 1990, + milder and wetter winters because of more westerly winds, + warmer and drier summers because of more easterly winds.

Source: KNMI.

The Netherlands will be confronted with precipitation that rises gradually year by year. More important, however, are the changes in the precipitation pattern, where a concentration effect will mean more short, heavy, local storms. This pattern causes higher peak loads, for which the urban water management processing capacity was not designed. At the same time, this capacity is being eroded by an increasing paved surface area within urban areas. Consequently, city residents are now regularly facing flooding, such as inundated basements and temporarily inaccessible parts of the urban infrastructure. The urban collection and processing capacity must therefore be enhanced to cope with the greater precipitation and the peak precipitation load.

The measures to be taken are drastic, necessitating an accurate specification of the nature and scale of the problem. Climate scenarios have therefore been developed for the Netherlands to clarify what we are facing in the future.

1.3 THE NETHERLANDS WATER LAND

The Dutch have a saying that 'God created the world, but the Dutch created Holland'. In Roman times, more than half the Netherlands was covered by peat, impassable and inaccessible. Considerable efforts were needed to make this country habitable and fit for urbanisation: the construction of dikes, dams and locks in combination with drainage. The Netherlands was transformed in the period between 1100 and 1700 from a natural and ever-changing system of lakes, marshes, streams and rivulets into an advanced water management system.

Improved pumping technology later enabled the draining of polders. The process involved pumping out shallow lakes, and the land reclaimed was at first given an agricultural function. The arrival of the water-pumping windmill and later the steam-pumping station boosted land reclamation. The land involved was generally a few metres below sea level. Urbanisation of the new land was hampered initially by poor drainage. The creation of an advanced water management system and the associated building of cities on the new land had to await the introduction in the mid-nineteenth century of the steam-pumping station.

Rapid urbanisation proceeded throughout Europe, especially towards the end of the nineteenth century. The urbanisation was accompanied by hygiene problems, which prompted the replacement of the open sewer system by an underground network of pipes, and at the same time a pipeline network for bringing in clean drinking water.

The Netherlands forms the delta of northwest Europe. Heavy rainfall in the 1990s caused flooding in the Netherlands and elsewhere in Europe because the drainage capacity of the polders and rivers was exceeded. At the beginning of this century, the carefully restored cultural heritage of Dresden ('the Florence of the north') came close to being engulfed by record high water in the river Elbe. The Netherlands is also threatened by water from the sea. Rising sea levels have started a new debate on the requirements to be set on coastal defences (Ministry of Transport, Public Works and Water Management 2004).

The advantage of the Netherlands over upstream countries is that it lies in a flat delta, so that much of the water that flows over the border can be divided effectively. There is plenty of space for water storage compared with upstream countries. Things went wrong in the Netherlands in the past usually when waters iced up, as can be seen on paintings of drifting ice destroying the dikes.

Most of the problems with flooding in Europe occur in the transitional areas between the mountains and the flatter regions, such as Prague and Dresden, where a river is forced through a relatively narrow passage, allowing river levels to rise sharply. If it rains heavily, the water will find a way through. Where cities have developed along the banks of a river, this water will end up in streets and buildings.

An additional difficulty in the Netherlands is that it is where the rivers meet the sea. The interaction can be dramatic.

1.4 THE ORGANISATION OF A WATER MANAGEMENT AUTHORITY

The farmers in the marshy fenland that was the west of the Netherlands in the early Middle Ages were obliged to drain the land by digging ditches, discharging the water into a river or artificial watercourse.

The population started to increase in the eleventh century, and much of what is now South Holland and West Utrecht was reclaimed at a rapid pace. This process demanded cooperation on the part of the land users, and the reclamation was usually a village affair.

This nascent water management authority was part of the everyday village business that was overseen by the village council. This arrangement worked well as long as the villagers were both owners and occupiers of their own land.

Differences of interest arose later, for example because the land was leased, involving people from outside the village. The drainage and flood defences that were necessary to keep the land usable were seen as a concern of the actual or beneficial owners. Because they were no longer the same as the village community, the village council could no longer represent the actual or beneficial owners, who started to form alliances of their own to protect their interests: water boards. These water boards were created on an ad hoc basis as water management problems arose, forming a patchwork of small areas, each with their own regulations. The actual and beneficial owners would enter into agreements among themselves for constructing a new dike or lock. When the next problem came along, they would make a new regulation, with yet another board. The government was sometimes involved in this process, especially if the intention was to implement a major project of interest to a large part of the country, such as the construction and maintenance of sea dikes. However, small-scale arrangements were largely a private matter, and would remain so until the nineteenth century.

A view that gained ground in the French period (1795–1813) was that the water management authority should be a public concern. After the French domination, in 1814, an educational programme based on the French model was set up for water management engineers, and the Ministry of Water Management was founded. The national water management agency was formed in the nineteenth century, and the provinces also had their own water management departments for the supervision and coordination of the countless small district water boards within their borders.

These small district water boards were gradually closed, combined, or divided among each other. This merger process was reinforced by the 1992 Water Boards Act, which not only determines that all the inhabitants of the Netherlands have an interest in and should therefore exercise control over and pay for proper water management, but also that all tasks to do with that management must reside with the district water boards.

1.4.1 Tasks and responsibilities of the Directorate-General for Public Works and Water Management

The Ministry of Transport, Public Works and Water Management works on the permanent protection against water and safe transport connections of international quality under its slogan 'Trusted with water, progressive in connections'. It creates innovative policy and is responsible for implementation and enforcement. Its most important concerns are the accessibility, safety and liveability of the Netherlands, with due regard to the responsibilities of the public, businesses and other authorities.

The two most important water-related units of the Ministry are the Directorate-General for Water Affairs and the Directorate-General for Public Works and Water Management. The Directorate-General for Water Affairs is responsible for flood protection and water quality. The Directorate-General for Public Works and Water Management directs the execution of national water policy, is responsible for the maintenance, management and construction of national waterways, flood protection, water quality and smooth and safe traffic mobility on the national waterway network.

The Ministry launched the public campaign 'Holland, living with water' in 2003, involving television and radio spots, posters, advertisements and the Internet. The objective was to raise awareness of the water problem in the Netherlands and the new water policy, which is set down in the government memorandum *Water policy in the 21st century*. Besides flood protection, another priority is water quality, because sufficient clean water is vital for people, animals and the environment. The Ministry therefore makes policy to guarantee the sustainable use of water. Daily water samples are also taken to check the state of the water quality. The principle that applies when problems are found is 'the polluter pays'. Because rivers do not stop at the borders, there is also substantial liaison with other countries on guaranteeing high water quality.

1.4.2 Tasks and responsibilities of the provinces

Together with the Directorate-General for Public Works and Water Management and the district water boards, provincial governments ensure that rivers, ditches and lakes remain navigable, that healthy swimming water is available, that domestic and industrial wastewater is treated, and that the groundwater is at the correct level. The climate changes are bringing more frequent and heavier rain. The policy has been adapted in anticipation of this change, and different water partners will be necessary. This partnership is embodied in the National Water Management Agreement (NBW).

The Netherlands is divided into seventeen regional catchment areas. The provincial governments, the municipalities and the district water boards have drawn up regional water plans for all catchment areas, known as the catchment area blueprints. They indicate where space exists for water storage areas and other practical measures, such as the restoration of streams. The provincial governments also draw up the catchment sub-area blueprints. The authorities are thus already creating space for water, without dumping the problems of excess water from one area onto another.

1.4.3 Tasks and responsibilities of the district water boards

The district water boards have traditionally built dikes and regulated the water level. Besides flood protection and the management of water quantity, the district water boards also attend to the quality of the surface water. Wastewater is carried through the sewer system to a sewage treatment plant run by the district water board. The district water board also removes storm water to prevent the flooding of buildings and farmland. Many plants and animals are able to live in streams and ditches, not least because the district water boards ensure clean surface water.

1.4.4 Tasks and responsibilities of the municipalities

The municipalities are responsible for ensuring good drainage and preventing flooding in the city. There is an elaborate system of pipes below the streets to collect wastewater. Both storm water and domestic and industrial sewage are collected and discharged into wastewater treatment plants. The municipalities manage the drains outside private property, which includes cleaning, inspecting, upgrading and replacing the sewers. The associated costs vary from one municipality to another, and are paid for from sewerage charges, property tax, or both.

The storm water problem is prompting municipalities to find solutions that retain the water in the city. The solutions are set down in a Water Plan, in which the municipalities present their view of how to manage water. Space for water is another important aspect alongside guaranteeing, or improving, the water quality and enhancing the subjective experience of water in the city.

1.5 GIVING SPACE INSTEAD OF HOLDING WATER BACK

In the past, the defence against water in the Netherlands has been based on the principle of holding it back 'for a while'. Coastal defences were improved drastically following the 1953 floods, and all dikes were raised in order to make a repetition less likely. In late

1993 and early 1995, the Netherlands was confronted with high water levels in the major rivers. Along the river Rhine 200,000 people had to be evacuated in 1995. The first proposals were to raise the dikes again, which would be expensive.

However, an insight that gained ground was that giving space to the water during peak loads might be a better alternative. The Netherlands has recently found itself with a surplus of agricultural land, partly because of European regulations. It is then far less expensive to allow the land concerned to flood occasionally. The principle behind the policy therefore became 'give water more space' instead of dike reinforcement. Emergency overflow areas were also designated. These are areas of land that can be flooded under controlled conditions in the event of high water. The primary objective of the measures is to retain, store and, only then, to discharge the water. The same principle is also being applied in developing strategies for handling the water in urban areas during periods of heavy rain: the aim is to retain and store.

1.6 INSTITUTIONAL POLICY FRAMEWORKS

Past measures in anticipation of sea level rise and climate change in the Netherlands have been based mainly on national policy (NBW). One result of the current broad public acceptance of the scale and urgency of the water problem is the European Water Framework Directive, which has legislated the ecology and the quality of water since late 2000. A Floods Directive has also been in effect since 2006. The Water Test has been introduced for urban areas.

The Framework Directive obliges European Member States to bring the ecology and the quality of the surface water up to standard. Action is required for each river basin. The Netherlands is divided into the four river sub-basins: Eems, Rhine, Maas and Schelde. The Framework Directive obliges Member States to draw up management plans for each river basin and sub-basin. The objectives of the Framework Directive must have been achieved by 2015, with a possible overrun to 2027.

A catchment sub-area blueprint has now been drawn up for all seventeen regional catchment areas into which the Netherlands is divided. The blueprint comprises an outline water task for the short term (2015) and the long term (2050). The water task specifies the measures to be taken for each sub-sector to create sufficient space for water. The river basin management plans (RBMP), detailing the measures to be taken, must be adopted in 2009.

The European Commission's Floods Directive obliges all EU Member States to identify all areas at risk of flooding. The flood prevention measures to be taken for each basin must be identified. The directive also prohibits any Member State action that might increase the probability of flooding in other Member States. This 'non-transfer principle' is extremely important for the Netherlands.

If the Directive comes into effect, including the non-transfer principle, then Germany, France and Belgium will not be allowed to raise their dikes indiscriminately to protect against rising water levels, because doing so would cause the surplus water to flow over the border. The countries will be obliged to liaise with the Netherlands on any raising of the dikes. The flood plans will give the Netherlands a clearer view of measures being taken upstream. The Netherlands will then be better able to estimate the maximum discharge capacity for Lobith and Eijsden.

The national Start Agreement on Water Policy for the 21st Century came into effect in 2001. The agreement is between the government, the Association of Provincial Authorities

(IPO), the Union of District Water Boards and the Association of Netherlands Municipalities (VNG). The National Water Management Agreement (NBW) followed two years later. One of the NBW's provisions is that municipalities must draw up urban water plans by the first half of 2006, which they are doing jointly with the district water boards and within the framework provided by the catchment sub-area blueprint.

The urban water plans address the quantitative and qualitative aspects of urban water management. The *Urban Water Plan Guideline* produced by the Association of Netherlands Municipalities (VNG) and the Association of Water Boards (UvW) proposes the following urban water plan structure (Te Lindert 2005).

- Water quantity. This part specifies the quantitative aspects of urban water management, with the flooding problem as the main issue, related to which are the urban groundwater problem and the performance of the sewer system.
- Water quality. This module illuminates the qualitative aspects of water management, based partly on the Water Framework Directive.
- Water in the built environment. How is water management to be planned and executed within the urban area, and what impact will this have on the municipal organisation?
- Urban groundwater. Groundwater level and its control are essential items in the water system. The permissible fluctuation in groundwater level in existing urban areas is often small.
- The relationship between the water system and the water chain (see the following section), in particular in relation to drainage. The policy for separate storm drainage is also addressed. The aim of this policy is to decouple rainwater drainage from the sewer network to form separate systems.
- The control and maintenance of all urban water management.

Since 2005, municipalities have been obliged to present their urban plans to the district water boards for a Water Test. The aim of the Water Test is to involve water in spatial plans and decisions from the outset, which demands consultation with the water manager at the earliest possible stage. It is therefore not a retrospective test, but one intended to ensure an early and active contribution from the water manager, tailored to each plan. An aim of the Water Test is to apply and implement the existing water management and spatial policy effectively, not to produce new policy.

The greatest benefit of the Water Test is the joint *commitment* at an early stage of the initiator and water manager, culminating in the water manager's recommendation and explicit consideration of the water aspects in the plan, preferably in a section on water.

1.7 URBAN WATER MANAGEMENT

Urban water management comprises a complex system of water flows, requiring a distinction between five 'types' of water:

- rainwater;
- drinking water;
- wastewater;
- groundwater;
- surface water.

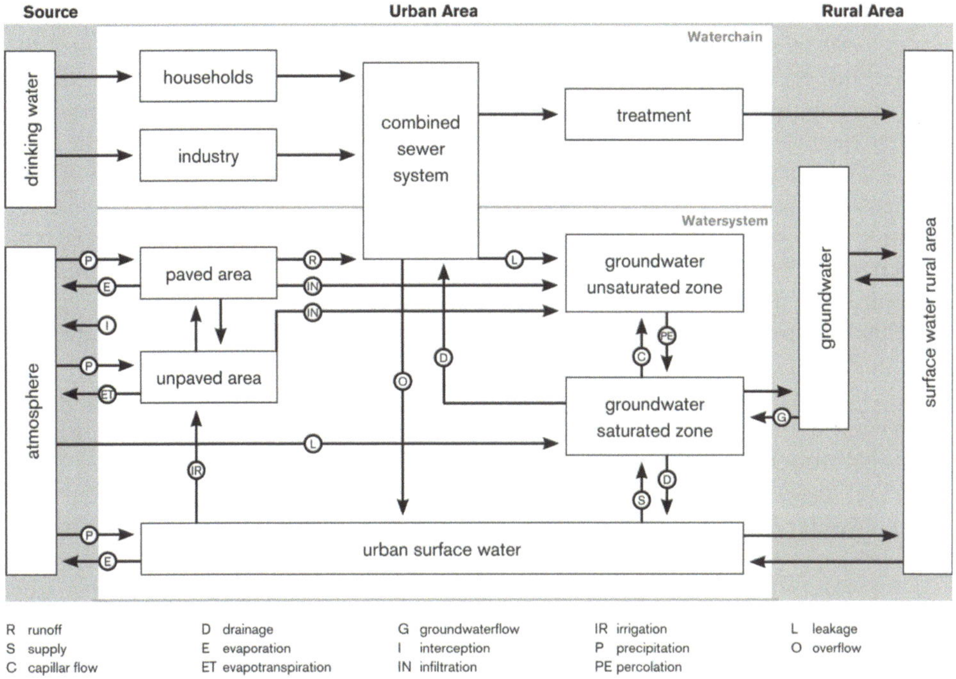

Figure 1.5 The urban water system.

Source: Frans van de Ven.

A distinction is made within urban water management between the water chain – which is the chain from the drinking water supply, through the sewers to the wastewater treatment plant – and the urban water system of surface water and groundwater. Urban water management involves two sources of water: the atmosphere and the drinking water supply. The wastewater treatment plant usually discharges its effluent outside the city.

A feature of the Dutch situation is that an average of 100 millimetres of water is let into the polders directly or indirectly from the major rivers that carry water all year long. This water is necessary for maintaining the correct level in watercourses. This is a fairly accurate affair in cities such as Rotterdam and Amsterdam, where the wooden foundation piles must never be allowed to dry out. The levels are often higher in the summer than in the winter. It is becoming more common to let water into urban areas, to allow the canals to be flushed through. The water quality deteriorates unacceptably if they are not flushed.

Urban surface water is usually separated from the surface water in rural areas, because of the more stringent requirements on level fluctuations and water quality in the city; an open connection could cause damage.

As much as 40% of the rainwater falling on a city flows away through the groundwater to surface water in the city and the surrounding areas. Only 23% flows away through the sewers, and the remaining 30% evaporates. The volume of water in the sewers is generally greatly overestimated, while much more flows away through the groundwater than most people realise.

The actual route taken by storm water depends on the type of sewer system. *Combined systems* were common in the past, in which household and industrial wastewater was mixed with storm water and transported to the wastewater treatment plant. The transport is subject to a limited rate of flow. During heavy rain, the storm water first accumulates in the pipes, pits and basins present in the mixed system, until they are full (equivalent to about 7 to 10 millimetres). At that point, wastewater mixed with churned up sewer sludge spills through *overflow pits* into the surface water – usually in the city.

Separate systems were introduced in the 1970s. The wastewater is then discharged through a *sanitary sewer* to the wastewater treatment plant and storm water is rapidly discharged through *storm drains* to the urban surface water, where it is stored before running off to rural areas. The *improved separate system* was added in the early 1990s. The process of disconnection and infiltration of storm water gained momentum in the late 1990s.

It is now widely acknowledged that relatively clean storm water does not belong in a wastewater treatment plant, but should be allowed either to flow away into the urban canals through a storm drain or surface run-off, or into the groundwater in the city. Various above ground and underground infiltration methods have been developed for groundwater. Infiltration fields, trenches, coffers, 'crates' and wadis are now widely applied, even on poorly permeable ground and in areas with high groundwater levels, albeit with sufficient drainage to allow the water to run slowly away to the urban canals.

The discharge capacity of storm drains is usually between 60 and 90 litres per second per hectare of paved area. Allowing for a certain additional storage provided by the street 'between the kerbs', the capacity is sufficient to prevent flooding. However, the discharge capacity of urban canals to the countryside is much more limited: nowadays it is about 12 to 14 millimetres a day, whereas previously it was 15 to 25 millimetres a day. Much storm water therefore has to be buffered in the city surface water and groundwater, which accounts for the substantial amount of open water in most cities. Whereas this was once about 3% to 6% of the area, it is now as high as about 6% to 10% in new neighbourhoods. A reclamation of 0.9 to 1.2 metres can then usually cope with a water level rise of 0.5 metres. However, a torrential downpour of 106 millimetres in 12 hours, like the one in the Delfland area in September 2001, and which happens less than once in 100 years, still causes limited flooding.

1.8 EXPANDING THE WATER STORAGE CAPACITY

The water entering the city therefore consists of rainwater and drinking water. The figures show the structure of the urban water mechanism and the measures needed to improve it and thus to limit the nuisance that water causes in cities. The following measures are discussed in Chapter 4:

- expanding the water storage capacity;
- using rainwater instead of drinking water;
- disconnecting storm drains from sanitary sewers.

It is an awkward business expanding the storage capacity in existing Dutch urban areas, but a need already exists for more storm water storage capacity. Separate sewer systems will be introduced in a larger part of the urban area in future, thus more storm water will be discharged into the surface water and there will be an even greater need for storage capacity. The following is a non-exhaustive summary of possible ways of enlarging urban storage capacity.

- The easiest and probably also the least expensive way is to temporarily lower the surface water level by preventive pumping. The surface water level is lowered by 10 to 20 centimetres when heavy rainfall is expected within a couple of hours. This measure creates storage capacity without requiring any spatial interventions.
- Additional surface water capacity can be created by widening existing canals and replacing decorative greenery by 'decorative water'. This option is supported by research into the use of urban space in the fifty largest municipalities in the Netherlands (G50), which points to the large proportion of public space in existing urban areas. The quantity of surface water should be roughly doubled relative to the current situation, in which approximately 4% to 6% of the urban area is surface water. Retaining water for longer will in future require an open water surface area of 10%. An alternative is to have inundation areas between 0.2 and 0.4 metres above the surface water level for peak storage.
- Historic city centres have a combination of little surface water and relatively little public space. Reopening canals that have been filled in presents an opportunity in these areas to create more surface water.
- Water can also be stored underground. Underground storm water basins can be developed by analogy with underground car parks. However, this would be an expensive solution that could be used only where no other, less expensive, options are available. A less expensive solution would be to use storm water as 'grey water' in buildings.
- Allowing the urban infrastructure to become submerged during heavy rainfall. This can be arranged by lowering parts of the infrastructure concerned. City squares are probably the most appropriate for this purpose. These 'water squares' would flood several times a year.

1.9 APPROACH ON A SPATIAL LEVEL

Seventy-five per cent of the existing property in Europe has been built since World War II. Particularly large numbers of buildings were built between 1950 and 1980. The property built in these decades is now technically and functionally outdated, as today's quality standards are higher. An important part of the building task for the coming decades will consist of restructuring and transforming existing urban areas. It would appear sensible to integrate tackling the water issue into upgrading operations of this kind. However, this integration runs up against the following problems.

- The usual aim of restructuring operations is to intensify the use of space, with a view to generating support for the necessary investment. Providing more space for surface water runs counter to intensifying the use of space for property. Creative design solutions are called for to unite these two conflicting wishes and to deliver synergy.
- Water managers are guided by the importance of effective water management in the districts and the regions. Project developers have a variety of objectives and

resources. The final plan for renewal and restructuring for both parties is therefore always the outcome of negotiation, in which each party will have been obliged to make concessions. The Water Test compels parties to enter into this negotiating process.

- The water system and the water chain are both part of an urban network. Objectives for improvement are set at an urban level first. Likewise, an assumption in the planning and execution phasing is that the realisation will be on an urban level. However, the realisation will have to be integrated into the restructuring or consolidation plans of the existing urban area, which implies that the improvement to the water system and the water chain will take place in small steps and spread over a large number of years, which demands considerable patience.

The background to spatial and functional interventions in existing urban areas is the discrepancy between the potential opportunities and the level of the current urban use. This means that there is scope for investments aimed at adapting the physical environment to the urban wishes. These adaptations balance the market position of the location, the urban area, and the function and status of the property in the area.

If the discrepancy between the current use and the modified situation is small, altering the existing buildings will often suffice. In other cases, the market situation offers more scope, and then alternatives can be considered, and a choice made between demolition and new building. This process is usually accompanied by an increase in the building volume and an increase in the building intensity. In an extreme variant, the urban development structure is also modified.

The functional changes that occur do not always have to be drastic, but could equally involve a change within a function. In residential areas, the old homes could be replaced by new ones, more in keeping with the standards of the time. For example, gallery flats could be replaced by single-family homes in a residential area. Old industrial estates could similarly be transformed into office sites, which actually involves a change in the employment structure. Sometimes a broadening of function also occurs, for example, when leisure facilities are added alongside retail facilities. Functional changes are therefore concerned with the extent of the change and the functional mix that then arises. It will be apparent that the complexity of the building task increases in line with the scale of the functional changes and the number of functions in an area.

The combination of a sharp rise in the use of space for residential purposes and developments aimed at greater concentration, and scale expansion in other sectors, has led to a segregation of urban functions. Social and economic trends would appear to be reinforcing this segregation rather than opposing it, which means that functional mix is a privilege enjoyed by few urban areas. Examples are the historic city centre and the modern city centre, where there has traditionally been a substantial functional mix. Another example is the district shopping centre where the residential function, leisure functions, office functions and perhaps also healthcare related facilities can be combined, through transformation and restructuring. A final example is the station site, where the retail function and office function can be added through transformation.

Spatial-functional segregation is a given. It usually manifests itself in interventions in the existing urban area as a heavy emphasis on mono-functional revitalisation, restructuring and transformation. The preservation of the existing land use and the urban structure has priority, although adapted to the standards and market conditions

Functional mutation			
Total function change	Functional revitalization	Functional restructuring	Functional transformation
Changing relationships functions	Multifunctional revitalization	Multifunctional restructuring	Multifunctional transformation
Function remains the same	Single functional revitalization	Single functional restructuring	Single functional transformation
	Preservation of physical elements and structure	**Balance between preservation and new structures**	**Emphasis on new physical elements and structure**

Figure 1.6 Typology of urban areas.

Source: Wout van der Toorn Vrijthoff.

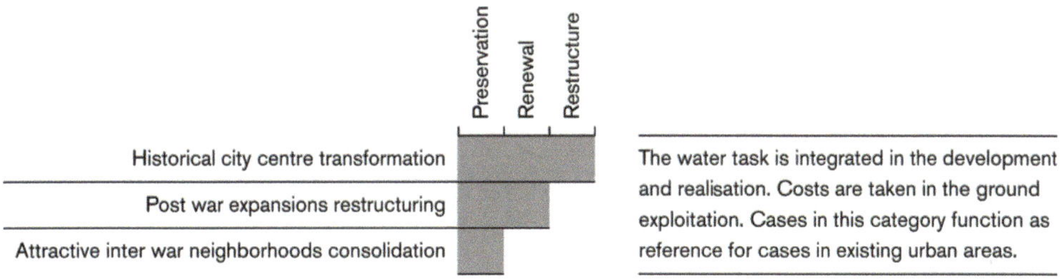

Figure 1.7 Typology of urban areas in relation to the water task.

Source: Wout van der Toorn Vrijthoff.

Figure 1.8 The waterfront of the city centre of Dordrecht.

Source: Fransje Hooimeijer.

of today. The districts concerned are usually large post-war residential areas, which have a relatively large amount of public space with ample greenery. Although the outline urban structure remains unaltered after restructuring, the part of the greenery with a decorative status is converted into water to satisfy the need for more surface water. The need to locate other functions in the historic city centre and modern city centre rules out application of this approach there.

The function change also applies to urban surface water. In the past 150 years, much urban open water has been filled in, especially in city centres, and for a variety of reasons. For instance, it was a way of providing more space for denser city centre traffic. The open water in city centres once had an important function for transporting goods. However, the city centre canals, harbours and rivers lost this function because of the scale expansion in inland shipping and the rise of alternatives, notably rail and later road. Lastly, the enormous expansion of the urban population, especially in city centres, has led to substantial pollution of open water, which was long used for discharging sewage. This was an added incentive to transport the stinking water in underground pipes.

The combination of a need for space, loss of function and pollution led to a considerable scaling down of open water in city centres, in particular in the twentieth century. However, there is a great need for more open water, in particular in city centres where the public space consists predominantly of hard materials. Coincidentally, at the start of the twenty-first century, the multifunctional use of space is becoming commonplace: functions are being stacked, both on land and on the water. Opportunities have arisen for creating more surface water, even in historic city centres. Conceivable solutions

Surface water	**Groundwater**	**Drinking water**
Discharge water surplus	Discharge water surplus	Human water supply
Store water (peak and seasonal)	Store water (peak and seasonal)	Household water supply
Supply water	Supply water to vegetation	Public health
Transport pollutants	Provide water (industry, households)	Industrial water supply
Break down of pollution	Transport pollution	Cleaning/flushing water supply
Retain pollution	Retain pollution	Irrigation water supply
Support aquatic ecosystem	Break down pollution	Fire fighting
Support terrestrial ecosystem	Support terrestrial ecosystem	Groundwater feeding (leak)
Water related recreation. e.g. • Boating • Fishing • Swimming • Strolling • Camping	Maintain anaerobic underground • Reduce weight (grainstress) • Reduce subsidence • Reduce oxidation (e.g. of peat)	
Urban landscape quality		
Separate (functions, areas)		
Cultural identity		
Housing (living boats, floating houses)		

Figure 1.9 Modern functions of the water system.

Source: Frans van de Ven.

include the restoration of canals, harbours and rivers that have been filled in, as well as smaller scale ideas, such as wadis.

1.10 COSTS AND SOURCES OF FUNDING

The restructuring of the existing urban area will take decades. It will therefore take just as long if not longer to expand the urban water storage capacity sufficiently in the form of additional surface water. The decoupling of the underground sewer network from storm water drainage can occur only when there is sufficient storage capacity on a spatial level for clean water above or below ground level. These components of the interventions for improving the urban water mechanisms are interdependent and must also be integrated in a stepwise fashion into the spatial restructuring.

Expanding the storage capacity and encouraging the domestic use of rainwater are not interdependent with the other improvement interventions. Domestic rainwater use is an improvement that pays for itself in each household. The necessary investments are balanced by lower costs, because less drinking water is consumed. The increase of the storage capacity serves a collective interest. The costs will therefore have to be covered by local collective charges. The costs per household will not be high.

Different cost settlement arrangements could be made for the interventions on an urban spatial level. The water manager would be advised to use the quality charge, because living by the water is considered attractive and waterside homes have a higher value (Bervaes and Vreke 2004). This benefit can be used to cover the costs involved in realising additional surface water.

Cost coverage of this kind can be applied in restructuring and transforming areas. Some of the existing property in these areas is often demolished, while other property is improved. The prices of the new and improved homes in the new situation are different from those of the original dwellings. The price difference also factors in the costs incurred in improving the water system.

The situation is different for consolidation areas. The homes in these areas – which might be attractive 1930s neighbourhoods – have a good market position and are in a sound technical condition. There is therefore no need for restructuring. However, the water problem in these areas still has to be solved. The associated necessary costs can be covered by local and regional charges that are not related directly to a specific area (see the figure below).

The above implies that one user will be charged more than another. The resident of a renovated or new dwelling in a restructuring area will pay more for a magnificent view of urban water. But this resident will also pay charges to the district water board

Higher real estate prizes (area specific)	**Area type**	**Higher taxes** (area generic)		
Real estate developer		Central government	Regional water boards	Municipality
	Transformation			
	Restructuring			
	Consolidation			

Figure 1.10 Who pays the bill?

Source: Wout van der Toorn Vrijthoff.

and the municipality. The average district water board charges in 2005, including purification charges, were 204 euros per household per year (*Atlas van de locale lasten* 2006). The average municipal taxes intended to cover the costs of construction, upkeep and upgrading of the local sewer system were 125 euros per household per year (Source: Association of Homeowners (VEH)). Although these amounts have increased annually above inflation, the price to be paid for dry feet and clean water is relatively low. Each household in the Netherlands uses approximately 92,000 litres of clean water a year, which amounts to 126 litres per person per day. The average drinking water price in 2005 was € 1.52 per 1000 litres (including VAT and mains water tax BoL). Almost 11% of the drinking water charge (€ 1.33 per cubic metre) is taxes (€ 0.15 per cubic metre) and 22.3% is the customer price (€ 0.34 per cubic metre).

An alternative financing system has been under discussion since 2005. Converting the current sewerage charges into a charge according to use could allow municipalities to spend money on disconnecting storm water and tackling groundwater problems.

The proposed legislative amendment on the basis and funding of municipal water management, also known as the Municipal Water Management Act, will have an impact on the Municipalities Act, the Water Management Act and the Environmental Management Act (Wmb). This Act will give the storm water policy a basis in the regulations. The amendments to the three Acts are as follows.

1. The Municipalities Act will be amended to give municipalities more scope for recovering the costs of municipal water management.
2. The Water Management Act imposes two duties of care on the municipality, one to do with rainwater and one for groundwater.
3. The Environmental Management Act (Wmb) will be amended on a number of points. For instance, several terms and the municipal duty of care arising from the Wmb (Article 10.30) will be clarified, and the municipality will be given the freedom to create byelaws on discharging rainwater run-off and groundwater. Furthermore, the scope of the municipal sewer plan (GRP) will be broadened to cover groundwater and rainwater, and a hierarchy will be established for dealing with wastewater.

Chapter 2

The form and function of water in the city

Fransje HOOIMEIJER[1] (and Govert GELDOF[2])

[1] *Department of Urban Design, Faculty of Architecture, Delft University of Technology, Delft, The Netherlands*
[2] *fomerly: Tauw, The Netherlands*

2.1 INTRODUCTION

The dynamics of the major water system (rivers and sea) and the regional water system (groundwater and rainwater in conjunction with surface water) have been vital for the reclamation and urbanisation of Dutch territory. Urban design was dominated in the seventeenth century, and again in the late nineteenth and early twentieth centuries, by striking a balance between technical efficiency and the intractable territory within a rational townscape. Modern technology was engaged in the post-war period to establish supremacy over the territory, giving the urban designer a blank sheet on which to project his design.

Figure 2.1 Mount town Hegebeintum in Friesland.

Source: Fransje Hooimeijer.

The balance struck between technical efficiency and the intractable territory has thus varied from one period to another, but an awareness of its cultural significance has grown steadily over time. Water structures are large-scale public works that have a structuring impact on the scale of the city (and are part of the system on a regional scale). They can acquire cultural benefit in details of materialisation and the relationship with the buildings.

This chapter sets out to highlight for each period (transformation, consolidation and restructuring) the logic behind the siting of the water structures in Dutch cities and how they relate to the urban fabric and the type of buildings. What changes have occurred in this relationship and what is the situation now?

2.2 BASIC TYPES OF WATER TOWN

The first steps on what is now the Netherlands were taken on the relatively high grounds, such as dunes and riverbed sedimentations, and later on artificial mounts, dikes and *burchts*. Living on mounts was a common practice mainly in Groningen and Friesland, and the building of burchts in Holland and Zeeland. The towns that were built on higher, possibly artificially raised, ground can be classified as geest, mount, river, coastal, burcht, dike and dam towns. The expansions of these towns were usually of a military nature, with fortifications and buildings inside polders: the polder towns. Water has a structuring role in all these interesting and unique Dutch town types.

Settlements expanded in the eighth and ninth centuries for military, and later economic, reasons. Towns, or, more accurately, villages, were created where there was an economic route or military boundary line. The Hansa towns of Deventer, Kampen and Stavoren and the centres in The West, Dordrecht and Tiel were built on the intersections of waterways. Amersfoort, Coevorden and Groningen developed at the intersection of lanes on dry land. The geests of Alkmaar and Haarlem, as well as Arnhem, Harderwijk and Rotterdam, were built where roads and waterways converged.

River towns, such as Maastricht, Nijmegen, Utrecht and Vechten, had mainly economic benefits from the water, and examples of towns that flourished because of their location on a military boundary were the burcht towns of Leiden, Middelburg, Oostburg, Den Burg, Doesburg and Breda.

Dikes and dams enabled increasing numbers of urban settlements to be created in the thirteenth and fourteenth centuries. A settlement would be granted the privileges of a town to reflect its economic and military importance and civilised culture, as evident in the production and import of food, a system of trade, means of defence, laws, rules, order and a political authority.

The choice of settlement site was a sensitive issue because of the threat from the sea and the rivers. It is striking that towns appeared precisely where building was hampered by poor ground or flood risk.

The spatial layout and situation of the water structure (or defensive structure in the form of a dike) was different for each of the types of water town mentioned above. The specific spatial characteristics involved are set out below on three levels for each type of town. The first level is the siting of the town within a larger water management system (e.g. a polder system, or a river system) and the conditions imposed by this water system then and now on the layout of the urban area. Secondly, the water system

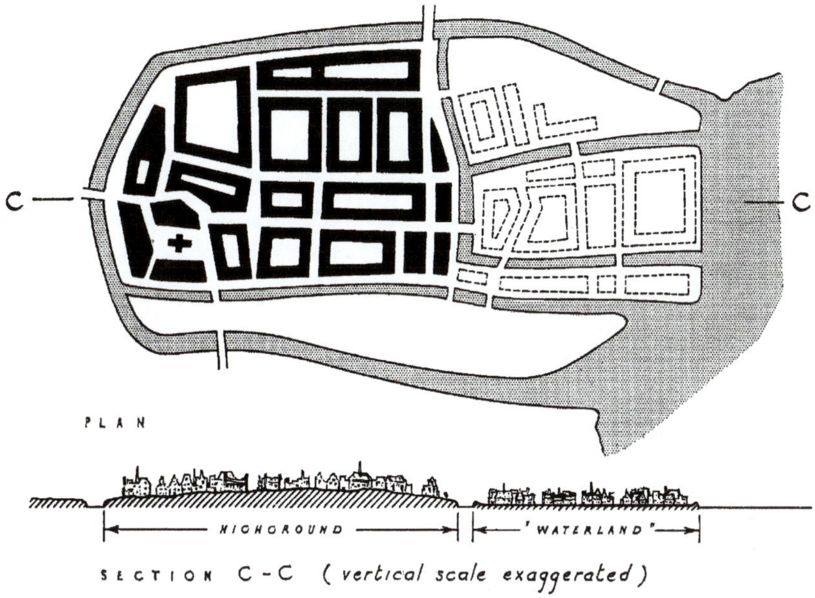

Figure 2.2 Geest town Alkmaar.

Source: Burke 1956.

in the town plan is the way in which a water system, or separate water elements (or water defence elements), form part of the structure of the town plan. The third and lowest level contains the water system elements and how they are integrated into the public space, including the technical approach and the details of waterfronts, quay, bank and water defence constructions. These features determine the potential of water elements as a valuable part of the urban public domain.

2.2.1 Geest town

Alkmaar and Haarlem are examples of towns that were founded on geests. A geest consists of strips of sandy ground interspersed with peat. They are found along the North Holland coast. An obvious advantage of living on a strip of sandy ground is that it is dry, but the surrounding peat land also formed an excellent defence against enemies. The sandy ground is part of the primary flood defence of the Netherlands, the North Sea coast. The elongated structure can still be found in the logic of these towns, in particular because the roads run along these sandy strips.

2.2.2 Mount town

The first habitation of the rough, wet Dutch landscape was facilitated by artificial mounts, especially in the province of Friesland. These residential mounts dotted across the landscape prevented individual houses or barns being washed away in the frequent floods in the time before dikes. Physical evidence of the expansion of mount dwellings to form a town survives in Leeuwarden and Dokkum.

Figure 2.3 Mount town Dokkum.

Source: Fransje Hooimeijer.

The elements of the water structure are visible in the different levels of streets and buildings. The level profile of the original mount pervades the architecture and the details of the buildings. The details of the expansion of the mount into the surrounding wet land with harbours and canals can also be identified in the other basic water town types. Canals, as the only public space, were lined on both sides with trees. The canals were also the only source of light, air and space in the compact town structure. The canal side building blocks were reserved for commercial use, industry and homes for the wealthier merchants. Workers' houses were built in the compact neighbourhoods behind, and there might also have been warehouses on the smaller canals, directly fronting the water, to facilitate loading and unloading. The design details of harbours and canals were concerned with quays (on which little information was available until the mid-nineteenth century) and of bridges and barriers. These structures were often effectively formed around the water structure, as an organic whole.

2.2.3 River town

Until the fourteenth century, the impetus for urbanisation in the Netherlands was a combination of economic opportunity, trade and power. A favourable location on a river that coincided with a trade route over land presented particularly good economic prospects. A market would then be established on the intersection, where goods would be traded. The first cities, such as Maastricht, Nijmegen, Utrecht, and Vechten, were therefore river towns (Rutte 2003). The differences in the dynamics and structure of the landscape created two types of river town, those on the levees and those against or on natural high ground.

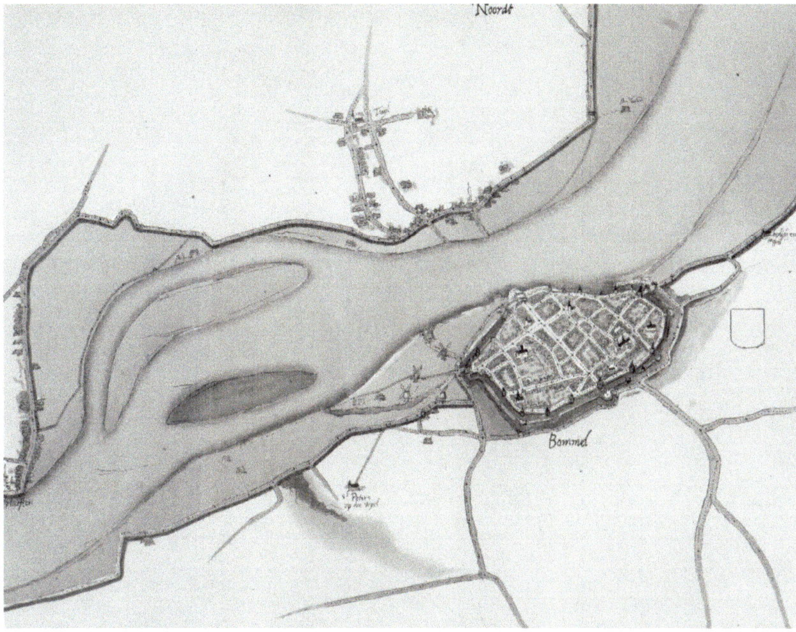

Figure 2.4 River town against natural high ground Zaltbommel.

Source: Hooimeijer *et al.* 2005.

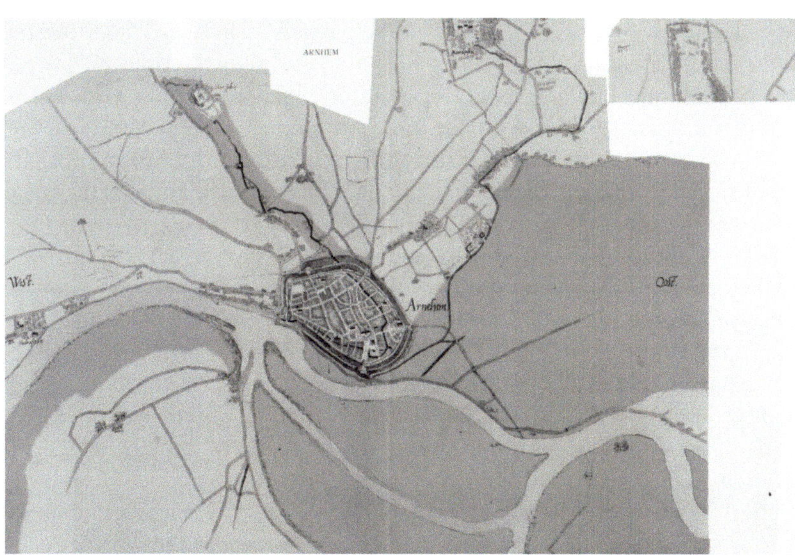

Figure 2.5 River town on natural high ground Arnhem.

Source: Hooimeijer *et al.* 2005.

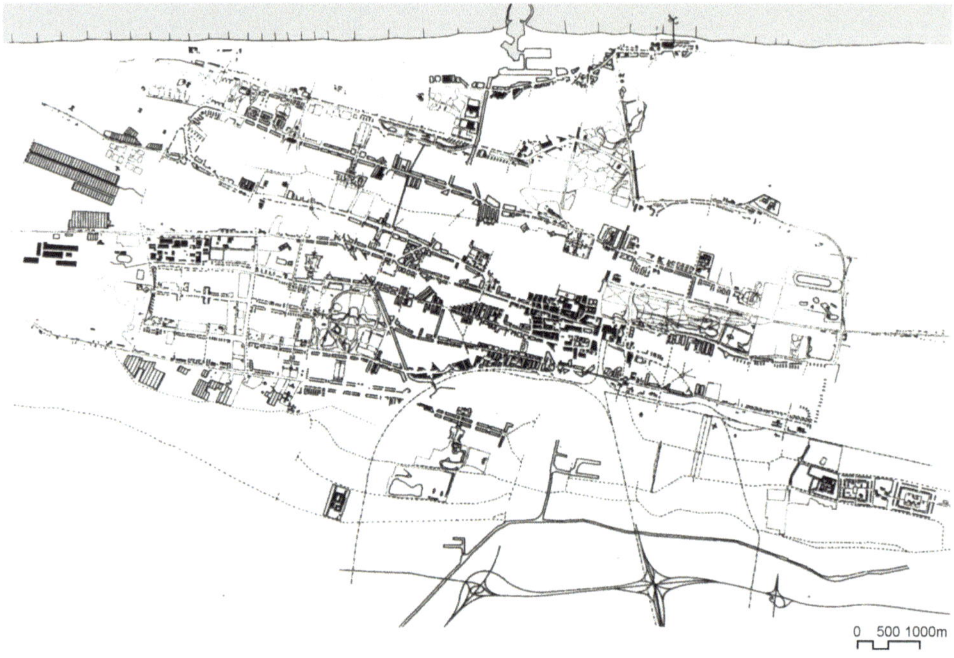

Figure 2.6 Coastal town on sand ridges The Hague.

Source: Hooimeijer *et al.* 2005.

The towns on the levees, for example Zaltbommel, formed along the contours of the land, which effectively thrusts itself upon the pattern of streets. The differences in level within the cities determined the risk of flooding, and thus the division of function and the style of building. Furthermore, the building blocks were aligned in various directions, as determined by the way the land was once parcelled out for agricultural settlement. Water was not extended into the structure of the urban fabric because of the risk of flooding (Burke 1956). The connections with the hinterland ran parallel to the river over the dikes, rather than along a tributary.

Examples of river towns against and on natural high ground are Maastricht, Nijmegen and Arnhem. These towns not only had a connection with the river, but also a strong relationship with the higher hinterland, with which they were linked by radial arterial roads. The important contact zones with the water system in most water towns were the quays, harbours and dikes, but there were none in the towns situated against natural high ground because of the distance to the water. The town square was usually too far from the river and had only the central function of market and meeting place. The differences in height produced an irregular urban pattern here too.

In a physical sense, the risk of flooding meant that towns were turned away from the rivers. This situation started to change in the seventeenth century, when more outer water was allowed into the town and the urban waters found a new role as urban decoration. Prestigious city waterfronts, such as the Boompjes in Rotterdam, became favourite business locations for leading merchants, exchanges and trading houses

(Meyer 2005). The 1874 Fortifications Act took away the military function and permitted IJssel towns such as Kampen, Deventer and Zutphen to build waterfronts facing the river. The cities along the river Waal, such as Zaltbommel and Tiel, and along the Lower Rhine and river Lek, such as Culemborg and Wageningen, remained with their 'backs' to the water (see: Hooimeijer *et al.* 2005).

The main sign of the detailed planning of the water system elements in river towns is in the solution to the differences in level with steps and sloping streets. Like the mount towns, this profile is also identifiable in the architecture. Temporary protective elements, such as groynes, have now also been made part of the public space, as has the furnishing with seats and viewpoints, which make the riverside so popular as a public space.

2.2.4 Coastal town

The coastal landscape has a wide variety of settlement types with numerous identifiable characteristics: dune villages, harbour towns, villages with the main street perpendicular to the dike, and key towns.

Dune villages were built safely in the shelter of the dunes, and in an economic sense were oriented to the sea. The oldest example is Domburg on the island of Walcheren, which was probably already part of the trade route to England in Roman times. The Hague is a dune village that was built further from the coast, but which still has evidence of the sand ridges in its urban pattern.

Building a harbour town along a sea dike required the dike to be built in a sort of loop, as in Goedereede and Veere. An outer harbour would be created, separated from the inner harbour by a lock. However, constant dredging was necessary to maintain the depth in the harbours. The urban pattern of the harbour town bore similarities with the river, dike and dam towns, in that the harbour fronts were reserved for offices

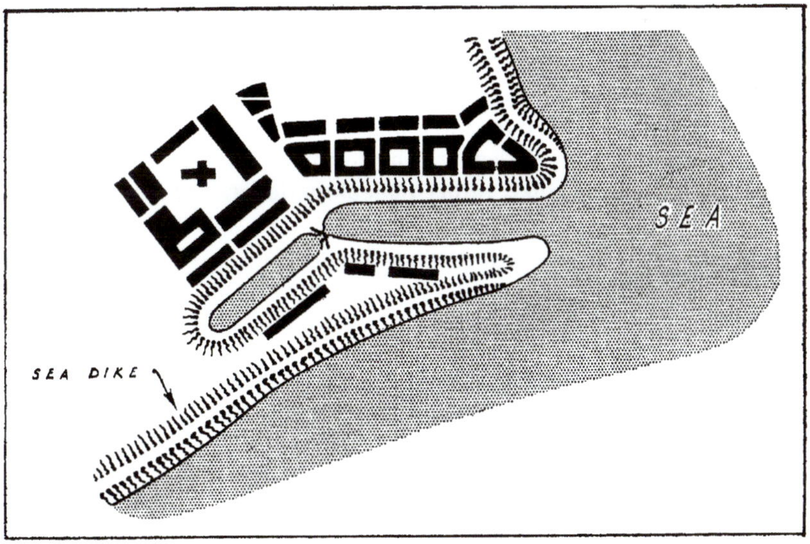

Figure 2.7 Harbor town Goedereede.

Source: Burke 1956.

Figure 2.8 Cape town Flushing.

Source: Hooimeijer *et al.* 2005.

and stores for merchants, boat builders and workshops, a weigh house and town crane, while the church, the town hall, the shops and houses were built on the lower ground behind the dike (Burke 1956, p. 24). Flushing's unique location at the deepest part of the Western Scheldt, which has changed little in the course of time, has meant that it is the only town to have escaped the silting up of its harbour.

The polder authorities enforced strict building regulations to avoid weakening the dikes. Building was prohibited on and alongside the sea dikes, which also left open the option of future raising or widening. The fishing communities would build a main street perpendicular to the dike, rather than alongside it. The fishing harbour in these villages with the main street perpendicular to the dike was outside the dike. The tree-lined main street was the site of prominent buildings.

The final type is the key town (the term 'key town' was coined in 1994 by Maurits de Hoog) which is a hybrid of the harbour town and the village with the main street perpendicular to the dike, where the harbour has silted up and become separated from the main water. A canal to deeper water was then dug to avoid the silting, and determined a new location for the harbour. The ground plans of towns such as Goes, Zierikzee and Middelharnis therefore have the shape of a key (see: Hooimeijer *et al.* 2005).

2.2.5 Burcht town

The spread of Christianity in the late ninth century was accompanied by the foundation of social and political institutions. Communities became more closely organised,

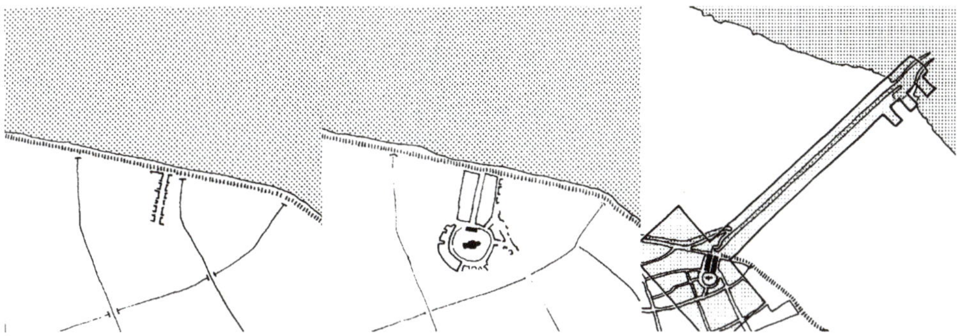

Figures 2.9.1, 2.9.2, 2.9.3 Middelharnis in 1550, 1650 and 1800 (key town).

Source: Hooimeijer *et al.* 2005.

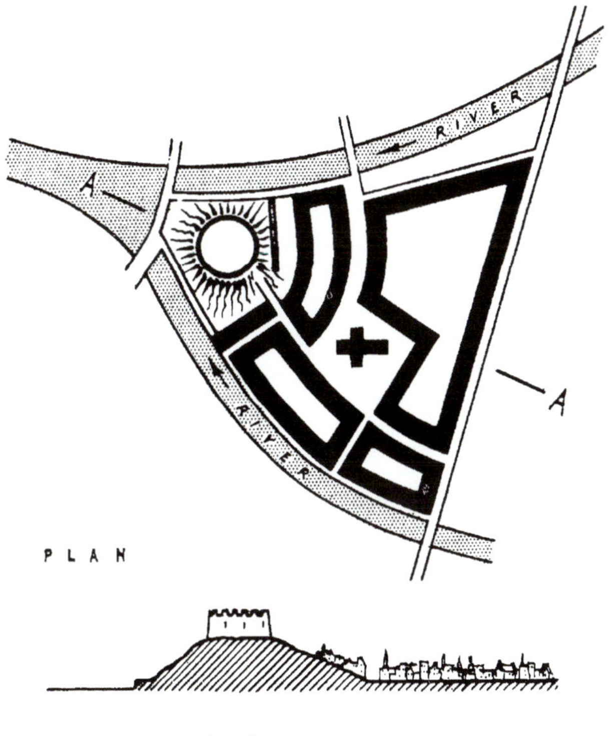

Figure 2.10 Burcht town Leiden.

Source: Burke 1956.

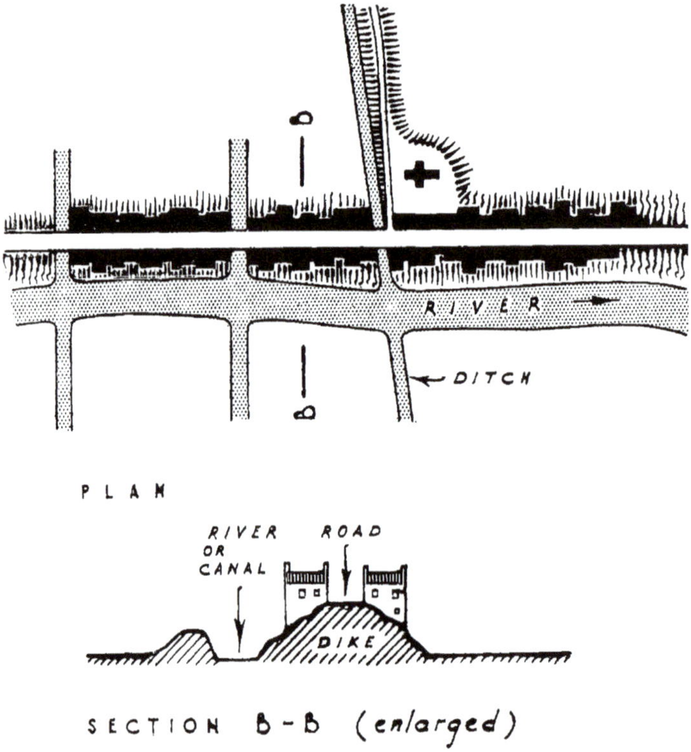

Figure 2.11 Dike town.

Source: Burke 1956.

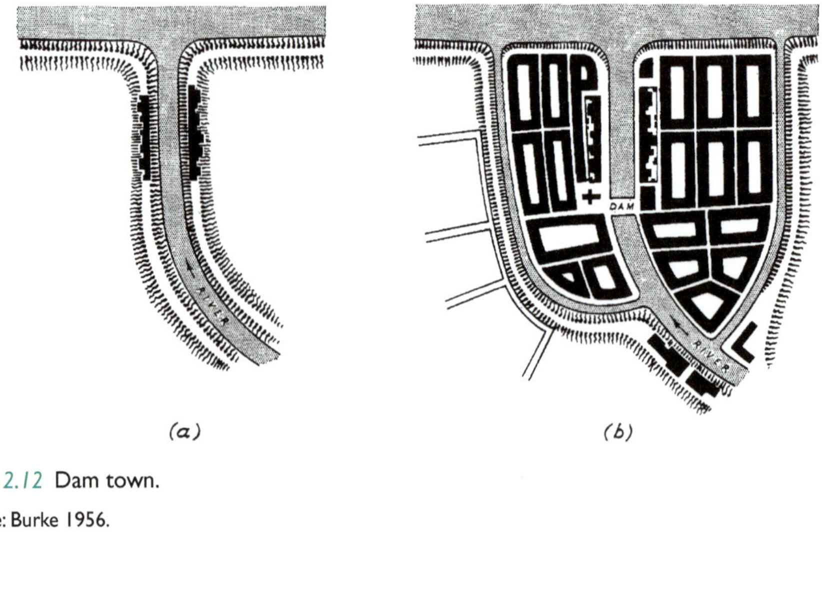

Figure 2.12 Dam town.

Source: Burke 1956.

and taxes and armies were introduced. Urbanisation, or collectivity, often occurred alongside or on a burcht, which was a round artificial hill that could be up to 60 metres across and 15 metres high, surrounded at first by palisades, then by a wall, and later a rampart with a moat. Examples are Middelburg, Oostburg, Den Burg, Doesburg, Breda, and, the only city where the burcht is still physically present, Leiden (Burke 1956, p. 23). The site of the burcht in this city at the confluence of three rivers, the Old Rhine, the New Rhine and the Mare, was most fortunate, both strategically and, in view of sedimentation, geographically. The central position of the burcht has survived all Leiden's waves of expansion. The siting the church directly behind the burcht and the expansion of a dike town along the three rivers confirm this spatially central position, as does the seventeenth century construction of a surrounding defensive wall and a moat.

2.2.6 Dike town and dam town

Dikes were built quite wide from the outset to allow a road and houses to be built along the top. Although the obvious function of the dike was to protect the dwellings in the low land behind, most settlements were also on the dike. Buildings directly against or on the dike were better protected against flooding and had direct access to a trade route. Other spatial characteristics were that the direct siting of the buildings on the water bound the dike town with the regional water system, and that the streets ran parallel to the watercourse.

This ribbon development still exists as a settlement form, but was also the basis for urbanisation. In times of expansion, the linear nature of the village was interrupted by placing a larger building, such as a church, behind the dike, so that a new square perpendicular to the dike gave the village a new centre. A centre of this kind also formed naturally in villages where a lane crossed the waterway. An example is Sloten, where the intersection is accentuated by the town hall, while the church was built alongside a tree-lined canal (Burke 1956, p. 37).

When a dam was constructed in a dike village at the point where a peat stream flowed into a larger water course, it became a dam town, such as Amsterdam and Rotterdam. The dike at these points was the most important condition for the creation of towns in the polders, because soil compaction and subsidence made these areas vulnerable to flooding. The dam had a water defence function, but with a drainage sluice, it also took care of discharging river water into open water. A combination of the scouring effect of the sluice water and the tidal movement were cleverly used to maintain the harbour at the correct depth and make the town accessible for seagoing ships. The economic significance of transport over water from sea to the hinterland and vice versa was embodied in the dam with drainage sluice, which became the heart of the city. The drainage sluice was able to accommodate only relatively small ships, and the cargo from larger ships had to be transferred or traded on the dam. The dam would become a market, and the peat river estuary outside the dike a sheltered harbour. The dam town and the polder were therefore bound closely together, not only hydraulically but also economically. This was manifest in a spatial sense by building the central social functions, such as the weigh house, the town hall and the church, on the dam. Although filling in work has taken away the original water structure of the dam in Amsterdam, the central spatial and social position of the dam is still visible today.

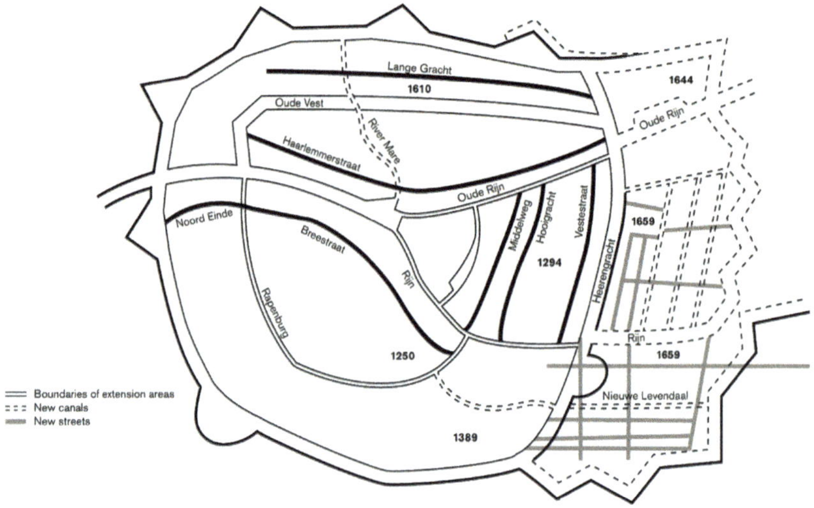

Figure 2.13 Expansion of Leiden.

Source: Burke 1956.

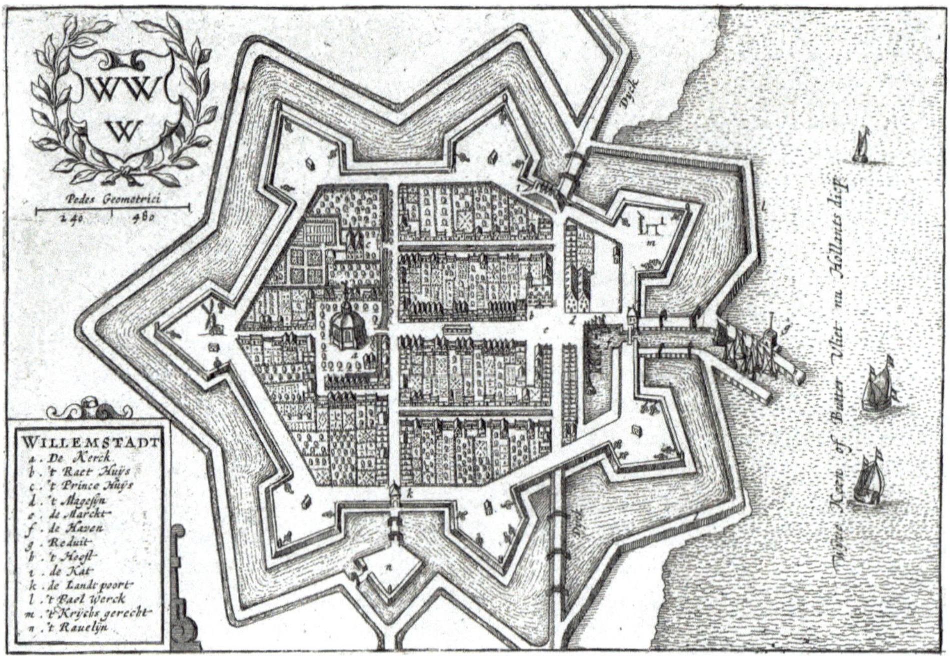

Figure 2.14 Willemstad.

Source: Burke 1956.

The difference between dike towns and dam towns in North Holland and Zeeland is that the type in North Holland had multiple canals, while the unpredictability of the sea in Zeeland meant that the water was linked with the town through only one canal that could be closed by a lock.

2.3 THE EXPANSION OF WATER TOWNS

From the thirteenth century, the privileges of a town were granted to the basic types of water town, geest, mount, river, coastal, burcht, dike and dam, on a large scale. The privileges included the right to hold a market, which was an important economic factor behind developing a settlement into a town. In view of the lack of any strong central authority, the towns had to defend themselves, usually by a wall. The economic system of town and surrounding areas, with the respective functions of market and production landscape, meant that it was in the town's interest to keep the surrounding areas dry. Similar to the way in which each individual town arranged its own defence, certain parts of the production landscape were ringed by dikes to protect against overflowing rivers.

Alongside the existing functions of drinking water, transport water, fire extinguishing water, storage and drainage, water in the thirteenth century also acquired a military purpose. The building of earth ramparts and moats around the towns was prompted in particular by the invention of gunpowder in the fifteenth century, against which a city wall provided no protection. Thus the fortified towns developed.

In the sixteenth century people started to learn to control water, or in other words to develop technologies to control the water management conditions. The overture to this period was the building of the first mill near Alkmaar in 1408. The mill could move a larger volume of water, and was therefore a more effective method of keeping the expensive city dry. Expansions in the polders depended on mills to keep the area dry. The polder town is therefore the second category of expansion of the basic types of water town.

It is obvious that the various types neither arose nor have been retained in a pure form, and almost all water towns are hybrids. Leiden, which developed through the three phases, is an example of a true hybrid among the water towns: from burcht town it changed into a dike town, and was later extended with a polder town. Likewise, a fortified town like Willemstad can also be a polder town.

2.3.1 Fortified towns

The fortified town is a later form of the burcht, based on the bastion, probably conceived around 1500 in Italy, as a response to the invention of gunpowder and the cannon. At the end of the sixteenth century, under pressure from the Spanish threat, Prince Maurits and Simon Stevin (1548–1620) developed urban defences based on the ideas and forms from the Italian renaissance. They are described in Stevin's book, *Sterctenbouwing*. Stevin was a great supporter of the use of readily available material, specifically earth and water, and developed a system with wide moats and earth ramparts. The material and the soil conditions imposed certain requirements on the design. He laid the foundations of the old Dutch system that the engineer Adriaen Anthonisz (1529–1609?) continued to develop in practice, as in Alkmaar (1574), Utrecht (1577), Naarden (1579), Coevorden (1580), Bourtange (1581) and Willemstad (1583–1585).

Ruigenhil was a village in the eponymous polder that was diked in 1564. Prince Willem I rebuilt it as a fortified town in 1583, because of its strategic position on the Hollands Diep. He called it Willemstad. The stronghold had seven corners; each named after one of the provinces, and had a church ring with a main street, which was a common type of town plan in South Holland and Zeeland. The street plan was influenced by the renaissance, and was formalised and modelled on the original polder land parcelling system.

The fortifications served as such until the French period, but fell into disrepair in many places in the nineteenth century. Besides the loss of the military function, another reason to level the ramparts was to use the space for hygienic functions, such as cemeteries and hospitals, but also as parks. The word boulevard is derived from the Dutch word *bolwerk*, reflecting the origin of these routes, now often used as promenades. Other motives were infrastructure, the construction of rail lines (Haarlem and Middelburg), the digging of canals (Middelburg), the construction of ring roads and urban expansion.

There were three ways of expanding a fortification. The first was to expand it around the town. The geometrical design of the surrounding walls was developed into a landscaping style and a relationship between the old and new towns was retained. Examples are Middelburg, Leeuwarden, Goes, Haarlem, Utrecht and Deventer. A second possibility was expansion around the town with a green ring. The rampart was retained, and the surrounding space became an urban park, as in Kampen. Finally, there were also fragmented expansions based on retaining the geometrical shape, with the objective of setting down the old city as a closed fortification in the open landscape, as in Naarden, Willemstad, Heusden, Elburg, Brielle and Hulst.

The water structures of fortified towns were generally around the outside, and canals crossed the town only for the occasional harbour or for drainage. The most important elements of the water system are the dikes, which determine the urban pattern together with the burcht and the fortification. For example, the sea dike in Willemstad is used for orientation of the siting of the fortification and is thus part of the urban design. The main street runs perpendicular to this dike, on which were the town hall and the arsenal, with drainage canals behind the housing blocks on either side. Again, elements of the water system are differences in level, fashioned with steps and sloping streets. Moreover, small bridges and pumping houses are included in the public space.

2.3.2 Polder towns

The above basic types of water town formed the first important spatial characteristic of the polder town: the higher ‘dry centres’ on which the settlements started. Prosperity and expansion led to expansion onto the surrounding watery soil, recovered from the peat, or already cultivated, but not yet suitable for building (Burke 1956, pp. 33 ff).

The most significant of the various dry centres on which peat polder towns were created was the dam town. The way in which the dike dwellers along a peat stream were able to control the water by building a dam was a precursor of the peat polder towns. The second spatial characteristic of polder towns was important in this regard: the necessity of ‘strict control’, which was a consequence of the painstaking nature of expanding the polder town.

First, the size of the expansion had to be determined. It had to satisfy not only the current needs, but also those of centuries to come. Second, a technical plan was required to ensure that the water could be discharged and controlled, and that the water level in the urban canals was kept stable. The initial work in most cases was to surround the expansion area with an encircling canal connected with a series of parallel canals running through the area. The encircling canal was intended primarily for drainage, but also had a military function (defence) and a transport function (access to the warehouses) (Burke 1956, pp. 64 ff). Locks and windmills regulated the water level in the canal system and discharged the surplus water. The land acquired would then have to be raised to achieve the desired level of protection, and be reinforced and prepared for building. The mud from digging the canals was used in the raising, for which it was mixed with soil that would often have been brought in from far away. Finally, long foundation piles would be driven into the ground cleared for building to provide a stable support for the dwellings on the deep sand layers.

The difference between the urban composition of the 'dry centre' and of the polder town can be seen in the ground plan of early medieval Alkmaar and its sixteenth century expansion in the peat (see Figure 2.2). The informality of the higher part built on a geest contrasts sharply with the strict control within the polder town. The difference in urban grounds levels is unmistakable: the streets slope gently away from the 'dry centre', while the water level in the canals remains the same.

The third spatial characteristic of the polder town, after the 'dry centre' and 'strict control', is the 'close involvement in the organisation and design of the polder landscape'. Peat polder towns were built on top of the agricultural pattern, often retaining it, while settlements in drained peat lakes were developed as part of the landscape. Examples of the two types are Delft and Hoofddorp, respectively. The structure of

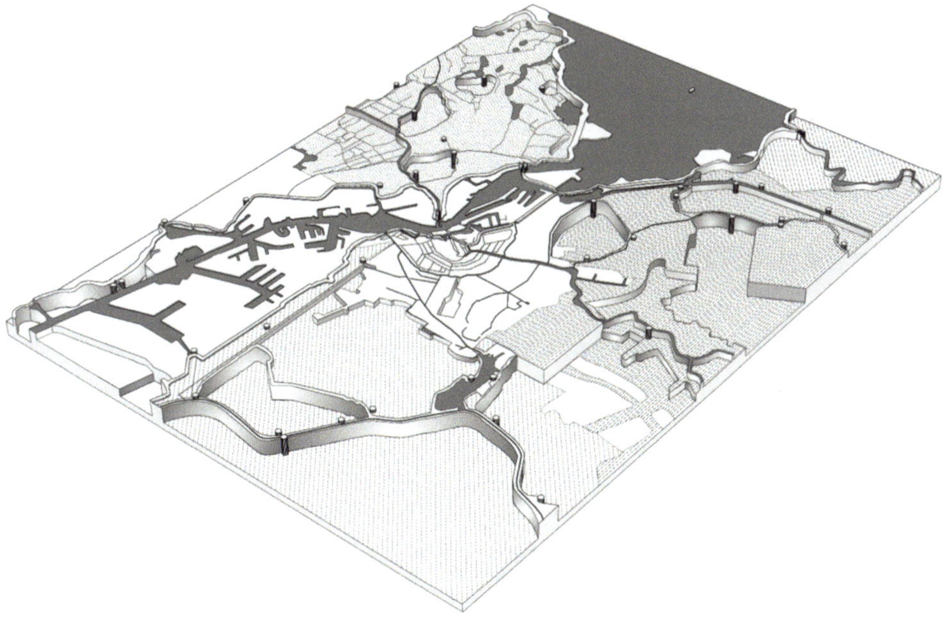

Figure 2.15 Polder town Amsterdam.

Source: B.B. Kwast.

Delft has a one-to-one relationship with the surrounding peat polder landscape. Hoofddorp expanded concentrically from the middle of the Haarlemmermeer polder, the intersection of the canal and the road, consistent with the dimensions of the polder land parcelling. Diked towns in the Zuyder Zee were developed together with, and independent from, the agricultural landscape, which explains the smaller scale of the geometrical relationships in peat polder towns, like the land parcelling in peat polders, than in drained lakes, and why diked towns exhibit independent patterns.

Two ways of dealing with water were devised in the polder towns: maintaining an outlet water level or creating a polder water level. A town in the peat had a choice between a polder water level and an outlet water level, depending on the existing level, whereas the only option for lower lying towns in drained lakes and behind diking (of the Zuyder Zee polders) was a polder water level.

The ground for a town with a polder water level is prepared for building by raising it slightly (by half a metre) and lowering the groundwater level. This means that the water from the polder must be pumped into the outlet waterway (*boezem*), to drain away naturally to a river, lake or the sea. This system requires a pumping station to be built, and more surface water is needed for water storage.

A town with an outlet water level maintains the natural drainage system, uses the outlet waterway for storage, and can therefore manage with less surface water because the storage is shifted outside the polder. Shifting water away in this way is no longer possible with the current water dynamics and because of subsidence throughout the west of the Netherlands. Depending on the original level, raising the ground level slightly may be sufficient for building an outlet waterway town, although a substantial rise (6 metres) may be necessary in some cases.

Technical progress at the end of the nineteenth century combined with burgeoning urbanisation put pressure on the polder towns. The hygiene problems caused by water in the town slowly started to influence the spatial impact of water management, aided by progress on the steam engine and later the combustion engine.

Construction of the sewer network and drinking water mains brought about a separation between systems for groundwater level control, wastewater discharge and drinking water supply. Much of the urban water system went below ground in the process. Furthermore, the train, tram and automobile started to displace traffic and transport over water at the end of the nineteenth century. This development prompted the filling in of numerous canals and waterways, with a consequent drastic reduction of surface water (De Vries 1996). The water structure continued to be important for storage and drainage in the polder town, but ceased to be used as a design instrument.

The elements of the water system (like the other types of water town) are the differences in level to be negotiated, bridges, quays and pumping stations. A significant new layout element in polder towns was the encircling canal. Because the main function of the water structure in polder towns was drainage and storage, there was no transport function and no need for hard surfaces, but rather a landscaped, green, park-like design.

2.4 THE OVERTURE TO THE CITY

The basic types of water town, extended with fortifications, polder, or both, retained their characteristic urban form until well into the nineteenth century. After the Golden

Age, in which most of these expansions took place, came a period of economic and political decline under the Spanish and later the French domination. The decline persisted until 1814, the founding of the monarchy. The period of urban development silence finally ended in 1850.

The French domination introduced an organised water management authority into the Netherlands, and a sense of water as a national interest. This idea was also expressed in water being seen as an economic driver, with King Willem I ordering the digging of many canals.

The first large-scale urban expansion was that of Rotterdam. In 1842, the city architect, W.N. Rose, developed an urban water system independent of the polder, known as the Water Project, which in modified form, with the help of the landscape architects J.D and L.P. Zocher, was adopted by the municipal council in 1854.

The plan served four objectives: flushing the urban water, lowering the groundwater level to allow building in the polder, a 'boulevard' and the creation of a residential environment for the wealthier citizens. The new polder town built between the

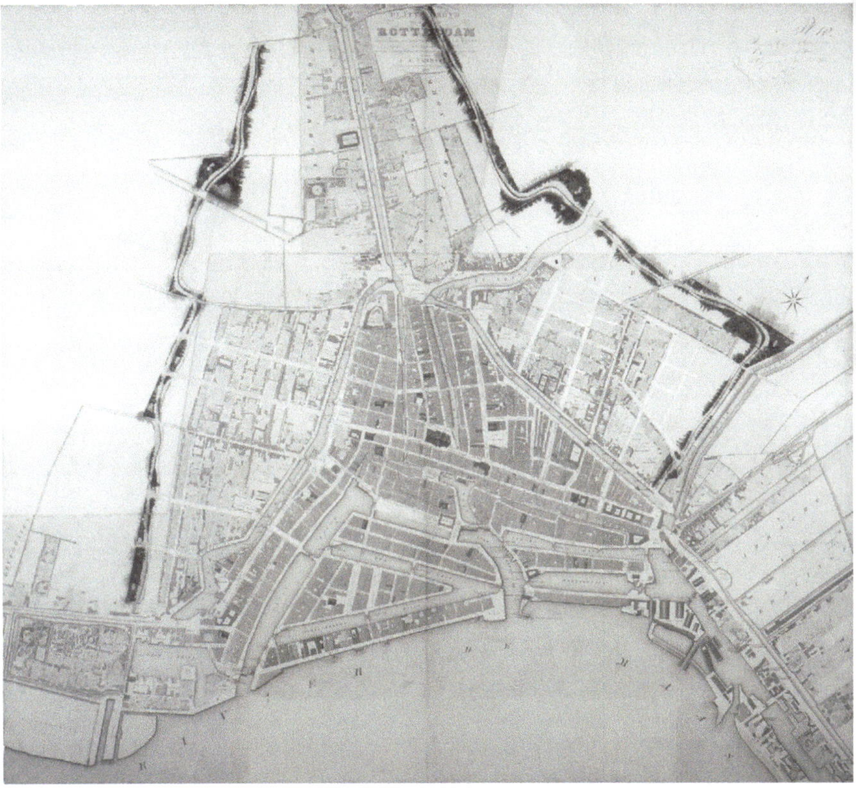

Figure 2.16 The Water Project 1854 for Rotterdam (see also color plate 1).

Source: Municipal Archive Rotterdam.

urban fortifications and the new encircling canal was completed within ten years, and Rotterdam repeated this method of expansion half a century later in 1910 in constructing the Heemraadssingel. However, the Water Project did not solve the problem of polluted urban water. Although the circulation was good, simply too much dirt was being thrown into the urban water, and the hygienic situation improved only with the construction of the sewers around 1890.

The 1874 Fortifications Act, which allowed the fortification works to be closed, was a long-awaited opportunity for many towns to expand. The Zocher family also developed an encircling canal project for Utrecht, in which the fortification was transformed into a green space for the city.

Amsterdam too was finally able to make the leap over the encircling canal (which had formed the urban boundary since the seventeenth century), but did not equal Rotterdam's conceptual calibre. In the nineteenth century liberal political climate, the city council left the expansion of Amsterdam largely to the private sector.

A nineteenth-century neighbourhood like De Pijp (the northern part of which was already being built before 1874) therefore lacked the elegant structure and systematic design of the earlier expansions. Like the earliest expansions that followed the old

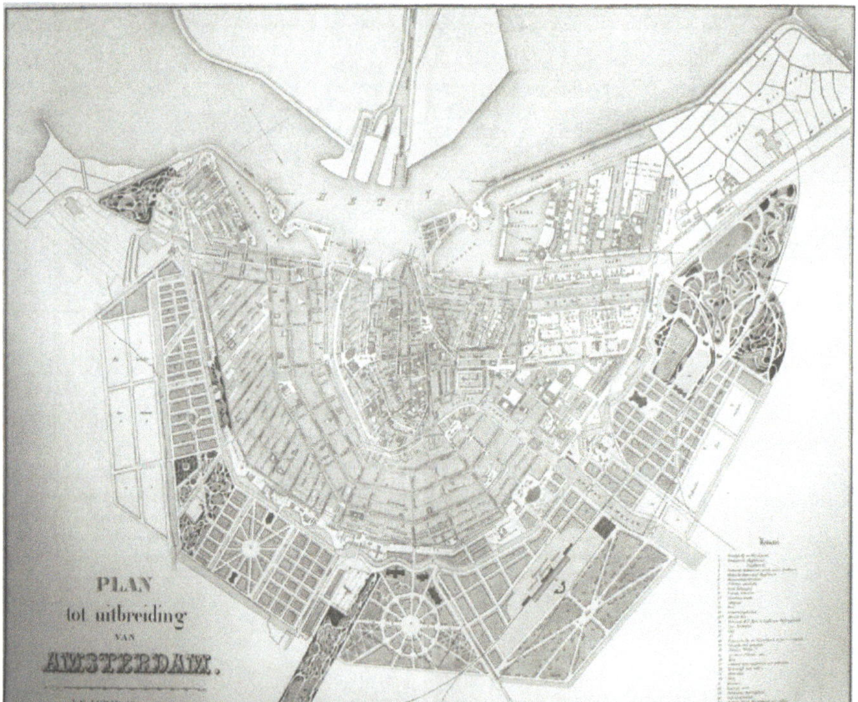

Figure 2.17 J.G. van Niftrik's expansion plan for Amsterdam (see also color plate 2).

Source: Municipal Archive Amsterdam.

pattern of grazing land and ditches, the nineteenth-century urban expansions conformed to existing structures, such as ditches and paths. The quality of the houses themselves also left something to be desired, and some houses even collapsed during the building. The phenomenon was referred to as 'revolution building'. The result was a ring of poorly built workers' neighbourhoods, which also caused problems with the water because of their small scale. The partial raising of the new building areas, in accordance with Amsterdam tradition to outlet water level, meant that the water flowed away to the intervening areas, where it remained, thus carrying the risk of malaria.

The water demanded a systematic expansion, which was one reason for the city engineer J.G. van Niftrik to draw an expansion plan as early as 1866. He projected a ring of new neighbourhoods around the existing city, separated by parks in the English landscape style. The plan has a formal structure with straight streets, star-shaped squares and rectangular building blocks with continuous façades and much greenery. The design is more oriented to architectural impression than the connection between the new neighbourhoods and with the existing city.

The unfavourable land parcelling involved would have led to massive compulsory purchase. The plan was duly superseded in 1875 by a pragmatic alternative based on the most lucrative land parcelling, produced by the Director of Public Works, J. Kalff. This design actually simply followed the polder land parcelling. This expansion plan produced new neighbourhoods such as the Staatsliedenbuurt, the Kinkerbuurt and the Dapperbuurt, packed with large numbers of inexpensive dwellings.

These poor residential building practices in particular led to the 1901 Housing Act. The obligation of towns with more than 25,000 residents to draw up an expansion plan prompted consideration of larger structures in the city. Unfortunately, this was not the case for the water structure, in which interest was waning because of its loss of function.

2.5 EXPANSIONS IN THE POLDER IN THE INTERWAR YEARS

The urban expansions in the interwar years were concerned principally with polder towns and were characterised by an interaction between the scale levels of the liveable neighbourhood and of the region. Around 1900 almost one-quarter of the Dutch population of 5.2 million people lived in the major cities of Amsterdam, Rotterdam, Utrecht and The Hague. As industrialisation advanced, industry and workers' housing were concentrated in the cities. Although the arrival of the car greatly eroded the function of urban water, people viewed canals, alongside streets and squares, as one of the three elements for shaping urban expansion.

The Bulletin of Acts and Decrees no. 158 of 22 June 1901 made important amendments to the provisions for expansion of the built-up area. The original component elements of the city were the street, the square and the park. Building was banned where these elements were to be situated. In 1901, the canal was substituted for the park from a conviction that digging canals had a higher priority than parks in the expansion of some cities, in particular Amsterdam.

Nonetheless, water lost ground to 'nature' in the city in the interwar years. The developments in the previous half century were of great influence on the building of cities in this period. The 1874 Fortifications Act allowed the cities to break open and

Figure 2.18 J. Kalff's expansion plan for Amsterdam (see also color plate 3).

Source: Municipal Archive Amsterdam.

to enlarge the overcrowded historic city centres. At first, the urban structures followed the agricultural patterns of the surrounding areas, but ideas on the relationship between the city and landscape started to take shape. For example, villa parks in the English landscape style were designed for wealthy citizens leaving the city.

The wholesale raising of expansion areas became less expensive because of advances in the technology for preparing building ground that accompanied industrialisation. Large-scale sand extraction from the inner dunes started around 1904. Nearly all the dune sand was originally used for preparing development land. Methods of estimating and new drainage techniques also had a major impact on the structuring role of water: it also lost its final meaning of drainage and storage.

The urban development plans from this period remained on paper because of the economic crisis. The urban development issue shifted to a regional scale, as illustrated by Albert Plesman's proposal in 1938 that give the ring of cities in the west of the Netherlands the collective name 'Randstad'. Amendments to the Housing Act in 1931 gave regional planning an official status. Growing mobility was putting pressure on the space in the city and surrounding areas. Sustained industrialisation demanded a separation of functions on a regional level. The Director of Local Works, W.G. Witteveen, developed a city plan for Rotterdam in 1926 moved to large-scale infrastructure, including ring roads.

Although this plan was not executed, its starting points are visible in the sub-plans that were realised, such as the one for Blijdorp of 1931. The thinking about the city was moved forward by various groups of architects, such as De Opbouw (1920), De 8 (1927) and the Congrès Internationaux d'Architecture Modern (CIAM 1928). Several CIAM conferences were dedicated to the 'functional city', in which the functions are spatially separated, with the objective of introducing light, air and space into the city. The style of building in continuous rows was introduced in order to align the buildings with the sun, for maximum incoming light. One product of these ideas was presented by Cornelis van Eesteren in 1934 in the Amsterdam General Expansion Plan (AUP).

Although water structures still played a functional and aesthetic role in the interwar years (e.g. in Plan South and in Blijdorp) the same was not true of existing water structures. Water structures continued to be filled in to hide bad smells, to solve traffic problems and for other spatial reasons.

2.5.1 Garden cities

The Garden City movement of Raymond Unwin and Ebenezer Howard around the turn of the century and Tony Garnier's design for a *cité industrielle* (1904) are examples of a quest for new concepts for the relationship between city and landscape. These ideas divide the thinking on the design into elaboration, construction, planting and building typology on the lowest scale level.

Raymond Unwin, in *Town Planning in Practice* published in 1906, urged a conscientious consideration of the existing landscape. This idea was implemented beautifully in Hampstead Garden Suburb, with its green centre and views from inside to outside. The principle was later applied in Frankfurt by Ernst May, with his urban quarters (*Siedlungen*) arranged along an open valley.

Influenced by Ebenezer Howard (*To-Morrow: a Peaceful Path to Real Reform* of 1898 and *Garden Cities of To-Morrow*) a socioeconomic ideal philosophy sprang up glamorising country living, contrasting it with the drab reality of industrial cities. Howard was concerned with the human scale and the enjoyment of air and space. He developed a garden city concept in which a large city was surrounded at some distance by a ring of satellite towns (with a maximum of 32,000 residents) having rural qualities and a higher level of welfare and happiness. The circular shape meant that they were not dynamic cities: they could not be extended. Tony Garnier countered with the idea of the *cité industrielle* (1904, published in 1917), a linear city, carefully situated in the landscape, and extendable.

These concepts also took root in the Netherlands, nourished by the uncontrolled spread of the cities and the rapidly deteriorating urban hygiene and environment. The large-scale building that started after breaking open the cities was scarcely supervised by the 'freelance' builders involved. The motive behind the sprawl was the builders' self-interest; there was no place for a communal quality of the city. The requirement for space for public greenery was reinforced around 1900 through the influence of health committees, which were founded to promote the hygienic interests of the city. Furthermore, the 1901 Housing Act was the first legal instrument for shaping the city in a controlled way, and facilitated the planning of greenery in the city.

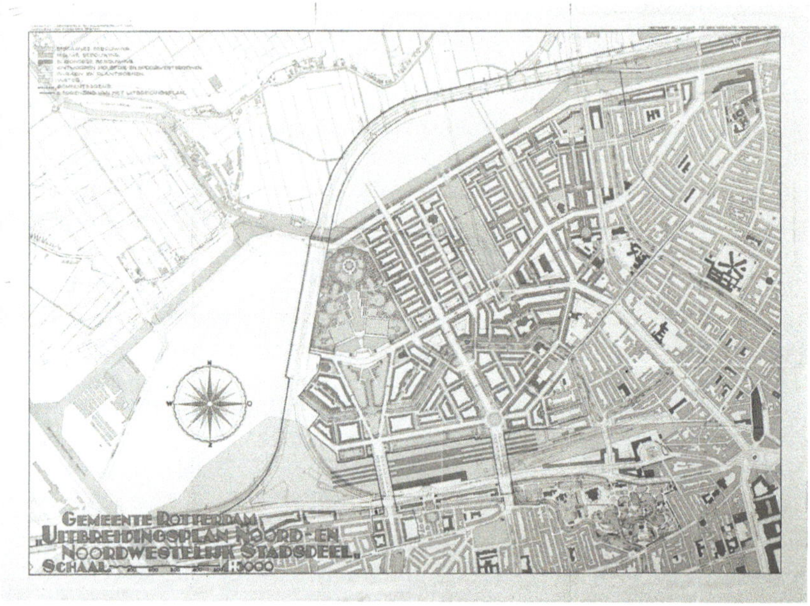

Figure 2.19 Expansion plan for Blijdorp (1931) (see also color plate 4).

Source: Municipal Archive Rotterdam.

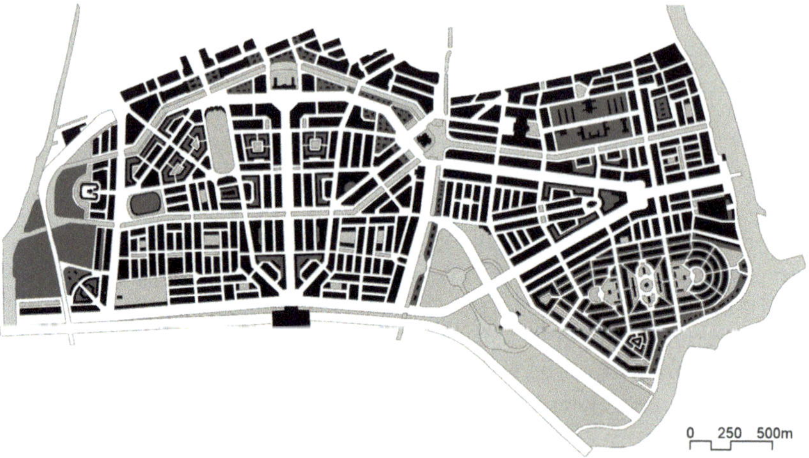

Figure 2.20 H.P. Berlage's 'Plan Zuid (Plan South)' for Amsterdam (1915).

Source: MaartenJan Hoekstra.

Alongside traffic, building and water, a new element therefore appeared in the urban structure: the public space. Simultaneously, the structures of building, traffic, and the combination of water and greenery were pulled apart from each other. These structures coincide in the traditional city. For instance, Amsterdam's *Grachtengordel* (ring of canals) has one main structure comprising all the elements. It would appear

Figure 2.21 'Plan Zuid' seen from the East.

Source: Municipal Archive Amsterdam.

that only together could water and greenery guarantee their claim to urban space at the start of the twentieth century. Water was part of the green structure, and the greenery acquired a right to exist as the exponent of water as public space. The composite structures of greenery and water originally retained some importance, and together with the traffic structure, they formed the backbone of urban design.

The unshackling of the water structure (with greenery) and the traffic structure progressed in the twentieth century. The structure of water and greenery in Plan South (Amsterdam) and in Blijdorp (Rotterdam) follows the traffic structure like a detached shadow. The building in both municipal districts accentuates the main structure in a traditional way. The style of building in continuous rows was applied sparingly, as in De Eendracht in Blijdorp (J.H. van den Broek).

2.5.2 Vreewijk

The First Rotterdam Garden Village Company was founded in 1913 by the prominent citizens K.P. van der Mandele, J. Mees and L.J.C.J. van Ravesteyn, for workers in the rapidly expanding Rotterdam. The company's purpose was 'to establish and operate one or more garden villages, especially for the less well-off class of the population'. The initiators were familiar with Ebenezer Howard and his *Garden Cities of To-Morrow* and wanted the garden village attributes, but not the isolated garden cities. The west part of the urban design of the urban expansion was developed by H.P. Berlage, and the east part by the firm of architects Granpré Molière, Verhagen and Kok. The urban structure of the garden village is formed by two long *singels* (gently winding waterways with trees), the Leede and the Lange Geer, and the Groenezoom avenue.

The original streams and ditches were used in constructing these *singels*, and the polder land parcelling is also recognisable in the urban pattern.

The first dwellings were completed in 1919, at which point what was literally and figuratively a halfway house between the countryside and the city became a fact. The aim throughout from the urban plan to the architecture and detailed design of the public space was to create an environment that belonged in the city yet fitted in seamlessly with the green character of the landscape. The street profiles were carefully proportioned, and the houses with their rural appearance and attention to sunlight (gardens at the time were mainly for growing vegetables), and the placement of trees, lanterns and fences were well thought out. The envisaged sense of a village was also underscored spatially by the construction of a brink, or village green, with numerous central services.

The garden village was extended repeatedly in the 1920s and 1930s, around a centre formed by the Leede. The streets became straighter and the homes less rural and more standardised, and the first flats started to appear. Vreewijk with its green attributes is still a popular working-class area, and has recently been largely renovated by combining some homes and executing various new construction projects.

2.5.3 Betondorp

The garden towns built in Amsterdam in the interwar years were Oostzaan (1920–1924), Watergraafsmeer (1923–1925) and Nieuwendam (1925–1927). Watergraafsmeer was

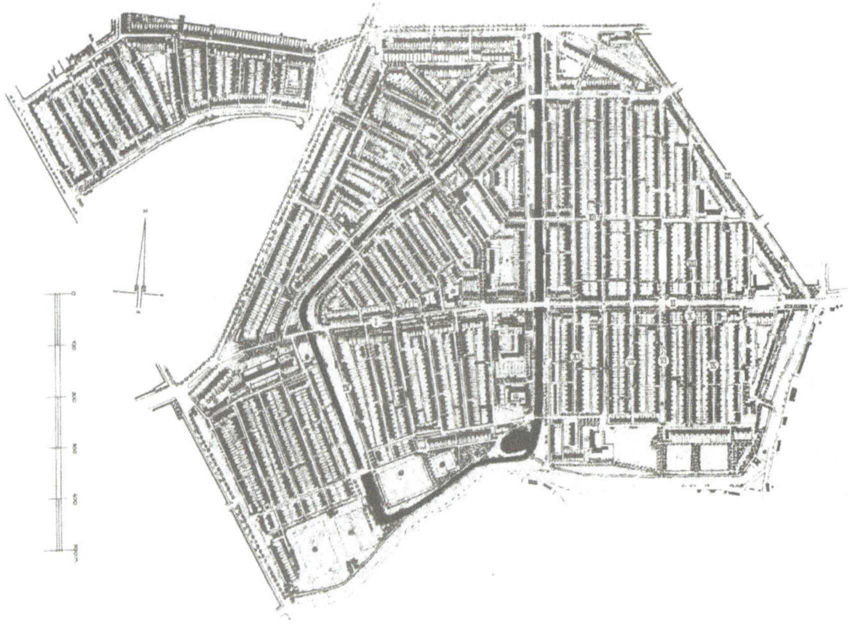

Figure 2.22 Plan for Vreewijk by H.P. Berlage and Granpré Molière, Verhagen and Kok (1911).

Source: Municipal Archive Rotterdam.

an independent municipality until 1 January 1921, built in the polder created by draining the Diemen lake in 1629. Watergraafsmeer is one of Amsterdam's lowest polders, where many well-to-do Amsterdam residents had country houses in the seventeenth and eighteenth centuries. The polder was laid out in a perfectly symmetrical 'wind form', with a grid cross determined by the dominant wind direction. The most important roads were the Middenweg and the Kruislaan, which divided the polder into four almost identical quadrants. The polder was surrounded by an outlet waterway, and in the middle, centred on the grid cross; it had a square pattern of ditches with connections running to the outlet waterway at the northeast and southwest. The way each quadrant was filled in meant that the water structure extended mainly along the edges. A *singel* was nonetheless located in the northwest quadrant.

The southeast part of Watergraafsmeer was named Betondorp (concrete village) because of the revolutionary concrete construction used there. High material prices and workers' wages in combination with a shortage of labour led to the cheap, rapid concrete construction method, which had already been applied in England. Nine hundred houses (400 single family and 500 two-family units) were built in the village, under the supervision of the Municipal Housing Department engineer A.F. Bakhoven Jr and the architects J.B. van Loghem and D. Greiner. Over one thousand brick houses were built alongside these concrete houses.

The village is shaped like a star around a village green with a public reading room and a clubhouse on either side of a housing complex for elderly people. The sense of a garden village is expressed in the village green and the many flower-filled front gardens. There were no fences along the boundary with the rear pool, but low privet hedges. The water structure was not used in structuring the plan, nor to impart a rural character. The only connection with the original ditch pattern was made along the west edge.

Figure 2.23 Betondorp.

Source: Municipal Archive Amsterdam.

2.6 WATER TOWNS AFTER THE WAR

World War II severely damaged built-up areas in the Netherlands. A government department for national planning was founded in the war, which was a forerunner of the Government Planning Authority. C. van Traa's post-war reconstruction plan for Rotterdam (1946), the reconstruction of Middelburg by P. Verhagen (1940), and of The Hague by W.M. Dudok (1946), are examples of a new approach to the 'open city' and the constant scale expansion.

After the war, population growth and inner city slum clearance boosted demand for homes, and large-scale expansion districts were built. The 'neighbourhood concept' and the open city philosophy were motivated by fears that the city would become too large to retain the social cohesion that could still be achieved in neighbourhoods and districts. The neighbourhood concept militates against the social disintegration and massiveness of the large city. A sense of security and a full community life were supposed to come about by putting the residents in clearly defined neighbourhoods. The open city idea relied on building in continuous rows and a spacious layout suitable for modern traffic. Both ideas were tried out in Pendrecht in Rotterdam (L. Stam-Beese 1953), and a mega-structure took the place of building in continuous rows in the Bijlmermeer (S. Nassuth 1973). The *woonerf* (home zone) was introduced in the Angelslo district of Emmen (N. de Boer and A.J.M. de Jong), and later the idea was much copied.

The disintegration of the various structures illustrates the separation of civil engineering and urban design (Van Eijk 2002). The designers of the *Grachtengordel* and the Water Project were military engineers and self-taught urban designers. Urban design emerged as an independent discipline in the early twentieth century, and the tasks were actually divided. The civil engineer solved the water problem and thus offered the urban developer an opportunity to develop a plan.

This opportunity was grasped enthusiastically, especially during the post-war building boom. The state of technology (better pumps and calculation methods) facilitated major ground area operations by hydraulic filling with sand. This and an underground drainage system meant that far less surface water was necessary.

Water has been viewed as a waste product since World War II, and has been relegated to district fringes and integrated into the infrastructure or the green system. The water system developed by the civil engineer is not visible as such, because surface water alternates with underground pipes. Furthermore, sand layers present the urban developer with a clean slate on which to set down any desired urban design, totally unhindered by the water system. Until 1940, the water in cities occupied between 12% and 15% of the total surface area, whereas in the post-war urban expansions the proportion was often cut to less than 5% (Meyer 2005).

The interweaving of networks has made way for a maximum separation of networks, according to both type and scale level. Each part of a network originally had its own function, which was not interchangeable, nor could it be integrated, with another network. A feature of modern post-war urban design, besides the much-maligned separation of function between housing, working, recreation and traffic, is the strict partition and distribution of tasks between the increasingly disjointed disciplines of urban design, landscape architecture and civil engineering.

2.6.1 Amsterdam: Western garden towns

Amsterdam's western garden towns resulted from the pre-war General Expansion Plan (AUP) for Amsterdam produced by Cornelis van Eesteren. Van Eesteren advocated a transparent city, in which wedges of landscape would both penetrate and blend with the city through a continuous fabric of green fields and apartment buildings. The careful orchestration of the various urban functions (housing, working, recreation and traffic) and the open and spacious layout has made the AUP into the icon of modernist urban design. It is the ideal of a pure, healthy, open and transparently ordered city, created through extensive urban survey.

Regarding water management, Van Eesteren conducted one such survey on the necessary and available quantity of sand. Amsterdam had been traditionally raised to outlet water level, and Van Eesteren had no intention of breaking with this tradition: 'The arrangement of the system of waterways, the water storage area, the system of sluices

Figure 2.24 Western part of the General Expansion Plan for Amsterdam by C. van Eesteren (1934).

Source: Municipal Archive Amsterdam.

and pumping, are all controlled through the level achieved by raising. A city built as an outlet waterway town has to be developed in an entirely different way from a polder town.' (Van Eesteren 1934, p. 159). Because: 'an urban expansion executed in this way [as a polder town] would have to be laid out to suit the pumping, one consequence of which would be a need to reserve a considerable surface area of the city for canals and watercourses to arrange for sufficient water storage.' (Van Eesteren 1934, p. 25), the lower polder town would have to be connected to an old city at outlet water level. Van Eesteren saw the connection of the road and water systems as particular design issues. The through roads would have to be linked by slopes, the canals by locks, and some scheme would have to be devised for refreshing the lower lying water.

The designs of the western garden towns after World War II exhibit a greater independence of the two elements of continuous building rows and traffic. These designs are also the first to include water in the public space. The original dike forms the main urban line in the design of Osdorp, with the traffic structure (which visually sets down lines to important buildings in the city centre) oriented perpendicularly. This rational treatment of the terrain as a basis for the town plan was facilitated by the hydraulic filling of the area. The raising to the traditional outlet water level as envisaged by Van Eesteren was cut back in 1950 because of the high costs involved. The western garden towns therefore have a polder water level and a far greater water surface area than foreseen in the original design.

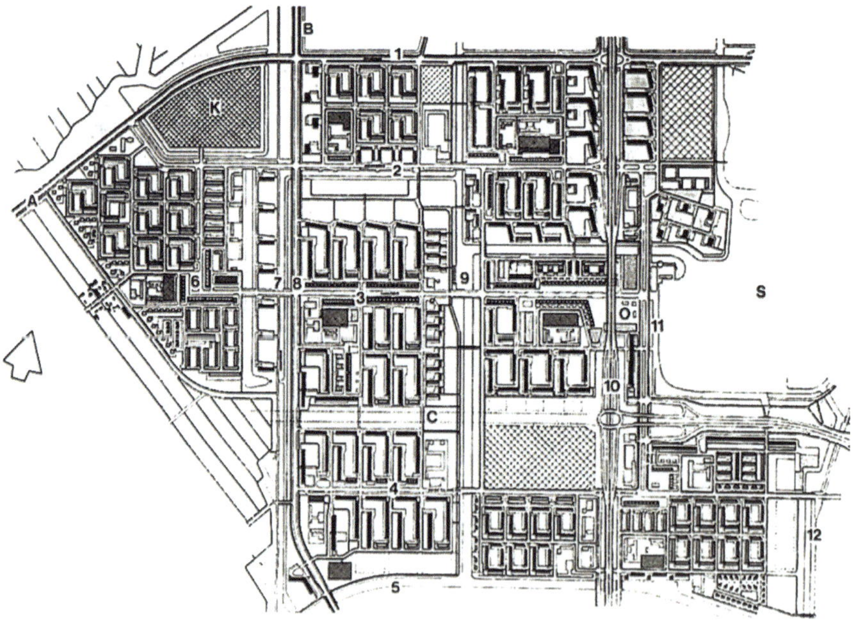

Figure 2.25 Osdorp.

Source: Municipal Archive Amsterdam.

2.6.2 Rotterdam: Southern garden towns

Pendrecht, Zuidwijk, Lombardijen and Groot IJsselmonde together form Rotterdam's southern garden towns. Like the garden towns of the interwar years, these districts were laid out in accordance with the garden city concept with much space and greenery. The main differences are the scale of the buildings and the layout of the public space. The post-war garden towns had many gallery flats and few of the houses associated with the foundation of the garden towns in the interwar years.

Rotterdam had a serious housing shortage after the war because of the loss of 25,000 homes. At the same time, there were forecasts of explosive population growth. The new districts would therefore have to include many homes, implying that high rise would be a dominant feature in the design.

The Lombardijen district was built in the 1960s in a former polder. The district was built in the shape of a wheel with a park and a shopping centre in the middle. The water structure surrounds the district and is a derivative of the original polder water system. Water has entered the district at several points in the form of a *singel* or a lake, but there is no structural use of the water as an urban design instrument.

The urban developer Lotte Stam-Beese developed Pendrecht, which acquired world fame for the progressive ideas on housing typology. The design of housing where families would have three or four bedrooms and a separate living room and, for larger families, a garden of their own, was revolutionary. Stam-Beese's urban development notion of open streets with much greenery where multiple population groups could

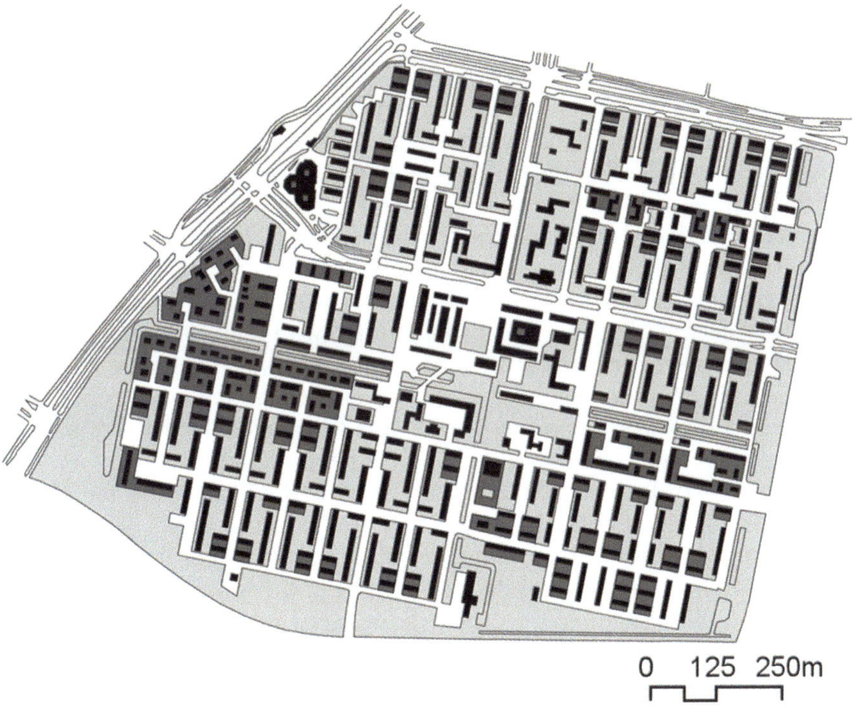

Figure 2.26 Pendrecht by L. Stam-Beese (1953).

Source: MaartenJan Hoekstra.

live alongside each other in returning building block configurations 'stempels' also attracted attention. The public space was laid out grandly and was intended for communal recreation and hanging out the washing. The residents of the time were viewed as a single user group consisting of equals. To make this feasible, there was even a selection process for residents who wanted to live in the model districts.

The water structure in Pendrecht also had little significance in structuring the town plan. Marginally, several stretches of *singels* were included, and one *singel* runs through the neighbourhood, along a road, through a park and past some back gardens. Nonetheless, water was a significant issue in the restructuring, and it was used as an element of spatial quality and of structuring value. The ideal community life, as portrayed at the time of building, and which was supposed to counter social disintegration, did not survive. The younger residents moved away, and the populations of the districts aged. However, inexpensive rented accommodation attracted many newcomers, resulting in a large-scale removal, which was ultimately to do the districts no good. Rotterdam's southern garden towns have now been in the spotlight for several years, and they are now attempting to shed their poor image. Various upgrade plans are on the table for the four districts. An interesting aspect of the plans for Pendrecht and Zuidwijk is that they emphatically continue the collectiveness theme, which was such an important starting point in the designs of these garden towns, and in which water also has a role.

2.7 THE HISTORIC DUTCH TALENT

In response to the technical approach to urban design of the 1950s, water acquired a special role in the 1970s in the aim of achieving cities on a human scale with an emphasis on *'gezelligheid'* (conviviality). The issue of 'urban identity' appeared on the political agenda, and led to the conclusion that water was an important component of the identity of the old Dutch towns. Since the 1970s, numerous plans have been developed and executed to restore earlier watercourses. For example, part of the old encircling canal in Utrecht was dug out again after having been filled in for traffic.

Unlike the austere canals of the post-war reconstruction, the natural character of a strip of greenery with water blended perfectly with the 1970s *woonerven* (home zones). Houses were arranged around courtyards and surrounded by green structures with *singels*. The natural character of water was used enthusiastically in the design of the water structure and the profile. The wildlife-friendly banks that became popular in this period altered the appearance of the *singels* considerably.

In the past two decades, water initially received little attention but later made a comeback in the city. Public spaces became bleaker in the 1980s because of the economic slump, which had a great influence on the application of the waterway as a natural element. The positioning of the public space in the urban design plan was purely functional, and the profile of the canal was designed for low maintenance. No financial resources were available for high-quality public space, and the design assumes minimum maintenance costs.

A conspicuous sign of the rising profile of cultural history and water management in spatial planning and design in the Netherlands is the reopening of historic canals, waterways, and harbours. Forty years after the last wave of filling in and influenced by

Figure 2.27 Pendrecht seen from the south-east.

Source: Municipal Archive Rotterdam.

the 'Belvedère' policy document (1999) and 'A different approach to water, water policy in the 21st century in the city' (2000), the restoration of urban water has become an important starting point in new spatial developments. Restored and existing canals and inland harbours provide an enormous impetus for the local economy and add value to the working and living environment of the trades people and residents of cities. There is also a three-pronged water threat (rainwater, seawater and river water), and cities that have plans for a dry future must find a spatial solution.

It can be seen from the account of the various types of water town that water infrastructure has considerable significance in forming the patterns in the town plan. The restoration of original water structures may represent an urban development and cultural improvement, provided it is carried out based on a correct interpretation of the meaning of water and with a sense of detail. Merely bringing the water back is insufficient. Urban water was carefully detailed in the past because the civil engineer responsible viewed the construction of a canal or other waterway as a total design. The sustained appreciation of this quality is crucial when cultural history is taken as the starting point for new developments.

In addition to the aesthetic and functional aspects in a physical sense, improving the system of public spaces and accentuating the leisure function can also be viewed in a different respect as history repeating itself (and therefore harking back to earlier cultural and historical values). The motive of many cities to restore historic water is likewise based mainly on economic benefits. Restoring water (and the ensuing maintenance) is an expensive and drastic enterprise, which would never be carried out purely for cultural and historical motives. Carrying out an urban ambition is the most powerful driving force behind plans, because less expensive and simpler instruments are conceivable for improving the public space. Perhaps another incentive for restoring water in the city is the three-pronged water problem. An important spatial implication of restoring water is that the expense and complexity involved forces a city to tackle numerous matters at the same time. It is precisely this cumulative effect of quality improvement that gives a new style to the ambition of urban revitalisation.

The tradition of coordinating the organisation, technical management and design of water systems has made the Dutch world famous. The fact that half the Dutch population live below sea level is the very reason for the existence of spatial feats, such as building magnificent polders and water structures for preparing the terrain for use. The function of urban water, the characteristics of the terrain and urban artefacts are responsible for fashioning unique patterns, demonstrating different and sometimes contradictory approaches to water. This tradition of combining the pragmatic with the aesthetic started to fade away in the interwar years, finally falling into disuse after World War II. The new attitude towards water is one of manipulation, of eliminating water as a problem and controlling it with technology. Water management was solved with civil engineering and vanished as a design instrument from the toolbox of the urban developer. The reintroduction of water into the urban development concept is not only necessary, it is also the reinstatement of a historic Dutch talent.

This talent may be given a better chance by scrutinising significant examples of the fine tradition in the light of current and future problems. The unreliable nature of the climate and the wilful behaviour of water and wind, which are responsible for regular flooding of the land, demand an innovative and less manipulative attitude towards water in spatial planning.

2.8 THE FUTURE: ROTTERDAM WATER CITY 2035

2.8.1 A leap in time

To conclude this historical summary, this section takes an exploratory leap into the future, based on the plan for Rotterdam Water City 2035. This plan attempts to unite the three mentioned water threats (river, sea and rain) with spatial quality and to make water an important vehicle for transformation and restructuring.

The municipality of Rotterdam and the district water boards produced a design for Rotterdam Water City 2035 in 2005. This design, portrayed in a scale-model, was exhibited at the second International Architecture Biennial in Rotterdam, which was organised around the theme 'The Flood'. The central question was how cities can deal with climate change and rising sea levels. Scale models of cities in the past, present and future were exhibited at the Biennial, showing how the cities are responding to and anticipating water threats and using water's positive attributes. A water system that performs well and healthily is a mirror for a good and healthy social system. Where society becomes muddied, so does the water, and its capacity to restore itself after

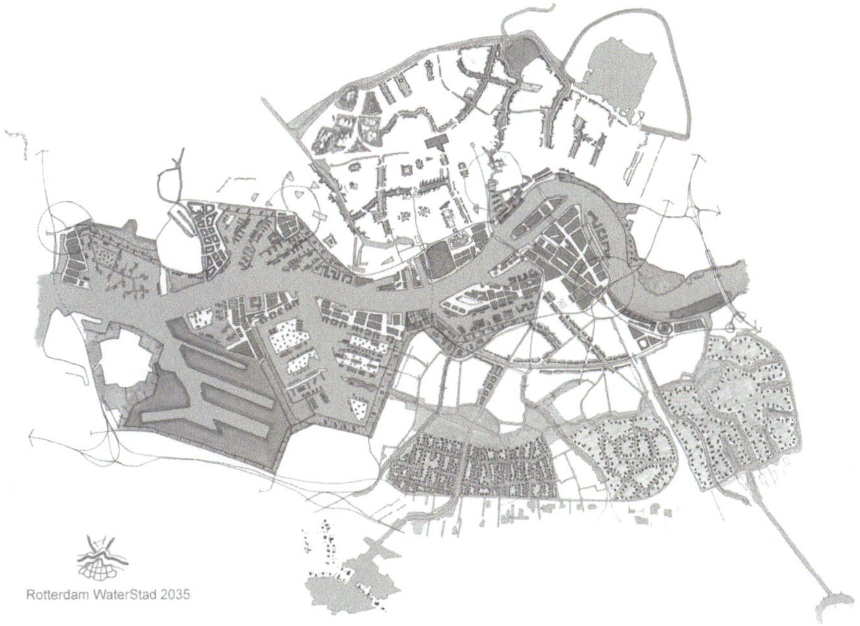

Figure 2.28 Rotterdam Waterstad 2035 (see also color plate 5).

Source: Municipality Rotterdam.

flooding declines. This idea is expressed in the design for Rotterdam Water City 2035 as follows: 'Water is our friend and enemy. We are providing answers to the hostile aspect by embracing the friendly aspect.'

2.8.2 The challenge

Rotterdam is a true dam town, which was built where the river Rotte emerged into the Nieuwe Maas River. A feature of the site is the accessibility of the city for ocean shipping. Its size was small and the opportunities for expansion were limited until the mid-nineteenth century. The poor soil conditions around the city and bad drainage locked Rotterdam in. Building was impossible in the wet peat areas surrounding Rotterdam. The city might have remained locked in for far longer had the Water Project of W.N. Rose not been executed. This plan provided for the construction of a *singel* structure and pumps, which improved the drainage and refreshed the water in the city. The Water Project's strength was that it transformed the threat into an opportunity. Rotterdam flourished after the execution of the Water Project, and has developed into one of the world's most important ports.

The city now again finds itself confronted with threats. Climate change and rising sea levels may eventually have an impact on Rotterdam's position. Socioeconomic processes could also enter a downward spiral, especially if the vulnerable position deters companies from locating in the city. Together with problems surrounding housing, urban decline, social segregation and a traffic network that is coming to a standstill, the result is a sombre future scenario. The challenge for Rotterdam Water City 2035 was to achieve a leap in quality regarding water. Like Rose's Water Project, the threats must be transformed into opportunities, which is the essence of the historic Dutch talent. A satisfactory response to climate change and rising sea levels will improve not only safety, but also the quality with which residents and companies are able to develop in the city. Rotterdam would thus become more attractive.

A group of approximately fifteen municipality and district water board staff members spent several months working on the challenge, in close collaboration with administrators, and have demonstrated that it is indeed possible to achieve the desired leap in quality. Gradually changing the urban fabric with water will create a city able to cope with extreme precipitation and water levels, while the water will enable it to integrate more values. The quality of life increases around water. Some of the aspects of the design for Rotterdam Water City 2035 are presented below.

2.8.3 The safety philosophy

One of the first questions to arise when presented with a challenge like the one facing Rotterdam is whether it makes sense to continue to live in the Dutch delta. Wouldn't it be better to move the activities from the low part of the Netherlands to higher ground? It is important to consider this question carefully, because Rotterdam's position in the low delta cannot be taken for granted. And if the safety philosophy remains unchanged, it might be untenable. Until recently, water safety in the Netherlands hinged on the height and strength of dikes and the ability to discharge water through watercourses, weirs, pumps and locks to the sea, often upwards. The safety was linked to the *probability* of these technical facilities failing. Safety was therefore a probabilistic factor, largely under the control of a relatively small group of engineers. The thinking was strongly based on the ability to make a living and working environment. The technology stemmed from weighing up the economic costs of the water defences against the benefits of being protected against water. The higher and stronger the dikes, the safer the Netherlands would be.

Rotterdam Water City 2035 was built upon the evolving changes in the safety philosophy in the Netherlands. It looked at the *risks* as opposed to probabilities, and showed more respect for natural forces. The feasibility concept has been retained, which is inevitable in a country like the Netherlands, but the idea that natural forces can be tamed completely has been abandoned. The combination of these changes results in an adaptive safety philosophy, a characteristic of which is that the changes implemented in the urban dynamics gradually make Rotterdam safer, so it will then incur less damage after heavy precipitation or extreme drought. Economics, safety and quality of life are thus linked together.

The concept of risk has two dimensions: threat and vulnerability. On the one hand, there is the water, of which there can be too much or too little. Its quality may also leave much to be desired. On the other hand, it is the urban functions that are sensitive to fluctuations in water quantity and water quality. Risks arise in the interaction between water and city. These risks are not constant, but are always in development. They are determined not only by technology, but also by social processes. It is possible to build upon the thinking of Ulrich Beck as expressed in his book on the *Risk Society*

(Beck 1993). Where people develop, they introduce risks. Risks are therefore a full part of our society. It nonetheless makes sense to keep risks as small as possible. The idea now is not to consider the technical aspect of risks exclusively – the reduction of threats – but also to consider the social aspects – people's vulnerability and activities in the city. The following methodology was used in Rotterdam Water City 2035.

First, extreme situations in water management were considered, to identify the technical facilities needed to improve matters. In other words, enough space must be provided to store the water produced by intense storms and to build up stocks of water to cope with dry periods with no unacceptable consequences for water quality. The required dike levels were also calculated. Consideration was then given to the urban dynamics and the characteristics of the existing urban fabric. Much more is at stake than water – indeed many Rotterdam residents consider the water threat to be less important than other problems – and how the various developments have an impact on the urban structure was investigated. There are traffic problems. There is a need for greater differentiation in residential housing. People are in search of identity and the Rotterdam economy is in a transition from a port economy to a knowledge economy. The city is changing even without a water issue. Rotterdam Water City 2035 looked at the opportunities for linking the desired technical improvements to all these developments. The water system that arises not only performs well under extreme conditions, but *every day is worthwhile* for Rotterdam and its residents. The ultimate result was new attitudes to traffic and transport in Rotterdam, differentiation in residential environments, economic performance and much more besides.

Rotterdam Water City 2035 adapts to climate change and rising sea levels and shows respect for natural forces. The approach is not one of maximum resistance to natural forces, but of bending with them as flexibly as possible. This process is referred to as enhancing the resilience of 'the Rotterdam system'. Therefore, there is no 'final situation' for the flood defences in 2035, but an adaptive strategy. The core is that we do not know what the situation will be in 2035. The fact is there are many uncertainties. What we do know is that we will know more about 2035 in 2030 than we do now. In other words, information will be added in the coming decades. It will be possible to make continuous use of the new information by designing a learning process involving experimentation and a series of 'no-regret' measures. Arranging safety effectively will be helped now that the port activities of the city are shifting towards the North Sea and the old port areas are being transformed into residential districts and industrial sites. Surveys have shown that Rotterdam can literally and figuratively keep its head above water until sea levels have risen by three metres, which is likely to take hundreds of years. If the level rises by more than three metres, then the strategy will change accordingly.

2.8.4 The design of Rotterdam Water City 2035

Anyone who knows what the city looks like now will notice some considerable differences in Rotterdam Water City 2035. There is more water in and around Rotterdam, with a significant variety of function and form. The water gives direction to the great diversity in residential environments in Rotterdam.

The centre of the city differs most strikingly from the current situation. It has been shifted towards the Nieuwe Maas River, following the important discovery that ground levels are above sea level along the river. The further from the river, the lower the ground level. If the zone along the river rises slightly with each transformation or restructuring, the safety will increase for both the zone itself and the polders behind. The design puts functions that are more public along the river – level with the current centre – and provides for 30,000 additional homes. These homes will be built either on new hills or piles, or will be able to float. The designs are adapted to the dynamics of sea and river. The river zone will become a highly dynamic urban zone that lives 24 hours a day.

The north of Rotterdam is largely earmarked for consolidation. The living environment in large areas is also highly valued, in the current situation. The changes in the area of water resemble surgical interventions in the urban fabric. An important aspect is the restoration of old *singels* in Rotterdam North. Numerous canals and *singels* were filled with rubble after World War II. Among them were the *singels* developed by Rose. The design sets out to reopen them where possible. The additional water area will be used for collecting excess water and maintaining the groundwater level in areas where the houses stand on wooden foundation piles. Where the above is infeasible, water squares will be constructed, which will be attractive spots in the city where people can meet each other and rainwater can be stored in the event of extreme precipitation.

Another important aspect in Rotterdam North is the reintroduction of the rivers Rotte and Schie into the cityscape; both are tributaries of the Nieuwe Maas. They are now somewhat hidden and offer little quality. New residential environments will be created along the Rotte and the Schie.

The south of Rotterdam will undergo a major transformation in the coming years, which offers opportunities for strengthening the living environment with water and improving the structure of the city. An important element of the design is that there is a single level for the surface water, which enables larger volumes of traffic and transport over water. An important design principle was 'everyone lives on the water'. A network of broad stately watercourses, connecting canals and residential *singels* add to the excitement of the area and provide a counterpoint to the pattern of roads and streets. The garden city principle will be revived in the southern part. Whereas there is a consolidation of homes in the river zone and in Rotterdam North, Rotterdam South will be thinned out, creating additional space for residents with average and higher incomes.

Much attention is given to traffic in the design of Rotterdam Water City 2035. It appeared possible to link the metro to a water-based public transport network with relatively few interventions. Rotterdam is already experimenting successfully with water taxis. This success will be extended in Rotterdam Water City 2035. While the metro mainly provides efficient north-south connections, waterbuses serve the east-west links. People can also travel on the Schie and the Rotte in the north and on the new water network in the south. This delivers a starting point for the approach to the congestion in and around the city.

2.8.5 The city and people

Rotterdam is in a transition from a port economy to a more varied economy, an important vehicle of which is the knowledge economy. Rotterdam Water City 2035 provides both a good investment climate for entrepreneurs and a greatly improved quality of the living and working environment. This is the intention. Taking adequate action with a view to climate change and rising sea levels increases investors' confidence in Rotterdam as a business location. The municipality is investing in quality, which encourages private investors to follow suit. Rotterdam is thus working gradually upward along the river, so that it retains its view on the water and the prospect of a healthy future.

It is important to note that the Rotterdam of the future is less introspective than now. Rotterdam is adapting not only to the climate, but also to developments in the social context, where water also has an important role. The city is opening up through the water and invites people to come together. This is a basis for creativity. Creativity – as the important building block for the knowledge economy – develops when people interact frequently and intensively. And as long as people exist, they will look for locations on the boundary between land and water. Rotterdam Water City 2035 has many locations of this kind to offer, and also has the necessary innovative capacity.

This may all sound extremely idealistic and perhaps there is a little too much stress on the values of water. However, it is not impossible. It will be possible to transform the water threats into opportunities by actually working in the coming decades on the comeback of water and linking with it desirable social and economic developments. Niels Bohr once said: 'Prediction is very difficult, especially about the future.' However, he overlooked the fact that much of what is achieved in complex systems can be viewed with hindsight as a self-fulfilling prophecy. Rotterdam Water City 2035 might illustrate this point well.

Chapter 3

The urban design issues in existing cities

Evelien BRANDES[1], John WESTRIK[1] and Bernadette JANSSEN[2]

[1] *Department of Urbanism, Faculty of Architecture, Delft University of Technology, Delft, The Netherlands*

[2] *Department of Spatial Planning, Municipality of Rotterdam, The Netherlands*

3.1 INTRODUCTION

People have always put much effort into the city. The existing urban area is changing constantly. Roadworks, hoardings around building sites, and undeveloped demolition areas; everyone will be familiar with these sights from their own environment or travels. Our cities are not finished, and the question is whether they ever will be. But

Figure 3.1 The city is never finished.

Source: Evelien Brandes.

what urban design issue is appearing on the horizon? This chapter discusses this question in two separate parts: one for the pre-war and the other for the post-war city.

3.2 THE STATE OF AFFAIRS IN THE PRE-WAR CITY

Things have never been so good in a sociospatial sense in Dutch cities, except, perhaps, in the Golden Age. Although the public image differs – because of integration problems and publicity on crime related issues – there has been a conspicuous upward trend recently. Whereas many city residents left in the 1960s, declaring the residential area unliveable, run-down and inaccessible, the city is now back in favour and alive.

The city has been dealt with rigorously under the heading of 'urban renewal', firstly in the 'inner city regeneration' projects of the 1970s and 1980s, which concentrated on the run-down residential districts. The era of the large-scale approach, where the cities were tackled in a way reminiscent of the Delta Plan for upgrading the country's flood defences, is now behind us. This operation is now in its final phase, with the districts of the third ring (dating from 1918 to 1940) now either complete or in progress.

Any homes remotely eligible in the regenerated cities are now being snapped up by residents of all sorts: young and old, rich and poor, Dutch nationals and ethnic minorities. There are even families, who might reasonably be expected to be looking for more space, greenery and quiet, who are now opting for life in the pre-war city.

Figure 3.2 Terraces in the harbour of Dordrecht.

Source: Evelien Brandes.

The cultural facilities and shops in the centres are flourishing. Every city has carried out alterations in the centre, whether major or minor, within the framework of the 'revitalisation policy'. Public spaces in many cities, such as squares and street profiles, have also been restored to order. Far fewer shops have been relegated to the periphery than in many foreign countries. City life is starting to regain momentum even in a street like the Grote Marktstraat in The Hague, where demolition works and the tram tunnel construction have long left large undeveloped areas. Some of the offices have also stayed behind in the old city, despite the trend for moving to the fringes, which have better motorway access.

The prosperity of the cities is remarkable compared with the malaise in the 1960s. The response of policy to the decline was demolition and major transport corridor projects, the primary objective of which was to restore access to the city. This was supposed to bring the city more into line with the requirements of the time and make it attractive for commercial activities. However, the measures backfired somewhat, and even made the cities worse, both economically and as a place to live. Many people opted to move out, maybe to a centre of urban growth, to the edge of the city, or further afield, to Brabant or the northernmost tip of North Holland.

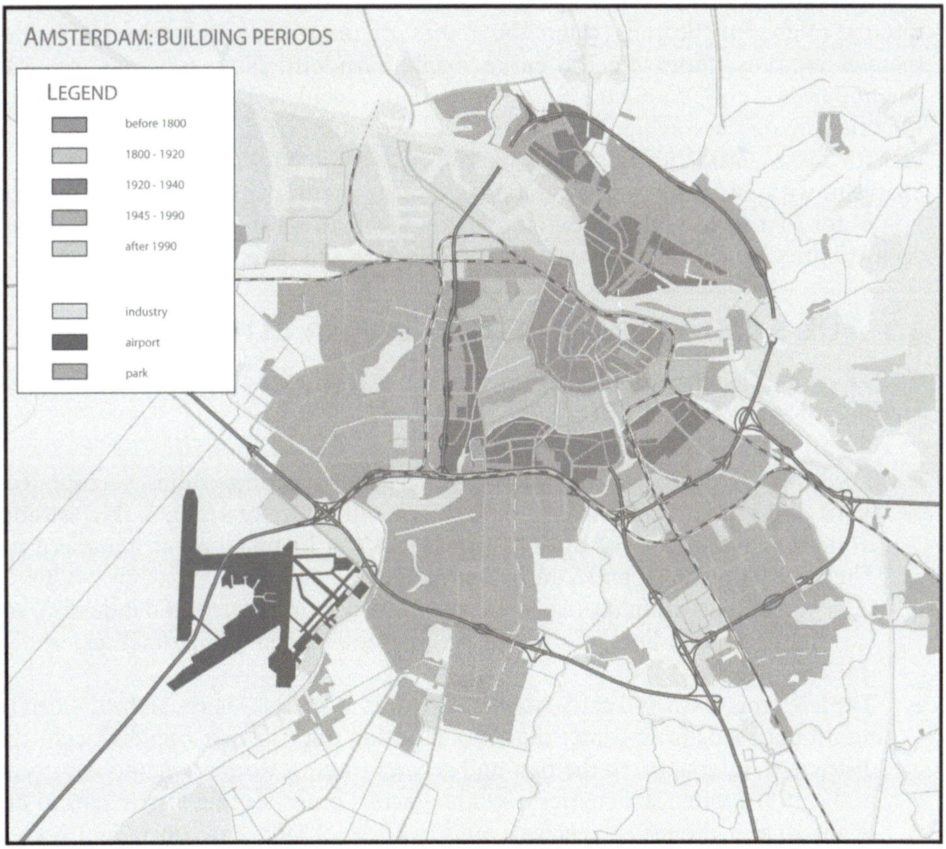

Figure 3.3 The historic structure of Amsterdam (see also color plate 6).

Source: D. van Veen.

There were attempts in the 1970s, under the slogan 'building for the neighbourhood', to renovate homes for the existing residents, especially those with lower incomes. This was as far as the city authorities were willing and able to go at the time. But even that turned out unexpectedly. The original residents still left the city in large numbers, to be replaced by immigrants in the old, renovated neighbourhoods.

The urban renewal of the 1980s did perhaps provide a quantity of dwellings to a higher building standard, but not sufficiently attractive residential districts. Furthermore, building was restricted to the social sector, for people with low incomes. Building for and investing in wealthier groups in the city were considered risky for many years. It was feared that the group would be uninterested in living in the city. Sporadic ventures for building apartments and town houses in the best city locations were limited to some high-rise riverside complexes or city centre sites. There were also some private initiatives for rebuilding and restoring special historic buildings and warehouses, or renovating old premises in the historic city centre or along a *singel* (a gently winding waterway with trees).

The urban area has now staged a comeback thanks to years of effort on more aesthetically pleasing plans with respect for the existing quality. Amsterdam's Jordaan and De Pijp districts are the well-known, relatively established examples of what is referred to as gentrification (where the homes are renovated by enterprising residents themselves, thus improving the entire neighbourhood). Now too, old, renovated neighbourhoods in West and East Amsterdam, such as De Baarsjes and Bos en Lommer, are on the way up. Rotterdam's Old West and North districts have been popular for some considerable time, but interest in living in once unattractive districts like Spangen, with an 'unliveable' stigma, has returned with the availability of attractive homes for first-time buyers. So now it is good to reap what has been sown, to grasp opportunities, or just to mark time for a while here and there.

3.3 THE STRUCTURE OF THE PRE-WAR CITY

The following elements can be identified in pre-war cities in the Netherlands (see also Heeling, Meyer, Westrik 2002):

- The historic city centre. This zone was once, and sometimes still is, surrounded by walls and ramparts. Most city centres are still intact or recognisable, although Rotterdam was destroyed by bombs in World War II and rebuilt in a modern style.
- The river with banks, parks and other green areas.
- Other locations within the urban area, such as station districts, old industrial sites, harbour areas, rail lines and junctions, and unintended 'fault zones' caused by a track or highway.
- The late nineteenth and early twentieth century districts. These can be divided into neighbourhoods from before the 1901 Housing Act (the first-ring districts – these districts were designated the first and second rings in Rotterdam and Amsterdam in the 1980s, often with extremely 'unhygienic' poor housing were changed drastically during the urban renewal), and districts of later date (including both the high-density second ring districts with many multi-family homes and the luxury 'villa parks' with a grandiose town plan). The residential districts concerned are therefore about one hundred years old.

Figure 3.4 The lively city centre of Amsterdam.
Source: Evelien Brandes.

- The expansions of the interwar years (1918–1940, the third ring). This is the period in which Berlage set the example in Amsterdam in the South and West districts. The districts have urban streets and squares in an imposing, often symmetrical, ground plan, in which garden villages and neighbourhoods with porch-accessed flats were built (Blijdorp in Rotterdam, and the Flower, Tree and Nut neighbourhoods in The Hague).

Amsterdam is a prime example of a city with a fairly accurately demarcated pre-war urban area, which is within the ring formed by the A10 highway. The ring road is combined with a railway ring, much of which runs next to the highway, with some sections running parallel somewhat further away. The ring is a clear marker. The true Amsterdam is inside the ring, where you still have a sense of being 'near the *Grachtengordel* (ring of canals)' and the cityscape is determined by long streets lined with high frontages and consistently closed blocks of buildings. Outside the ring is where the post-war urban area begins, with its more open, transparent and anonymous character.

The boundary between the pre-war city and post-war expansions is harder to discern in many other cities. The pre-war urban area is usually less well defined and blends inconspicuously into the post-war expansions, because the highway network lies outside the urban area, or, as in Rotterdam and The Hague, a motorway is actually situated inside the pre-war city.

3.3.1 City centre: narrow streets versus large-scale functions

We have come to realise in the recent past of constant urbanisation and globalisation that the Dutch historic city centre is unique. The Dutch seventeenth-century city represents as much symbolic value as acclaimed Italian cities such as Venice, Florence and Lucca. The proportion of historic urban area is declining, certainly in a world in which major cities spring up and are altered to suit contemporary building styles. It is therefore advisable to be more prudent with everything that has survived for more than a hundred years; not as museum pieces, not because they are old, but because they have qualities that cannot be repeated. A historic city gives an identity to the entire city and provides an attractive setting for people to gather and for tourism. Flourishing bars, cultural establishments, restaurants and pavement cafés are more likely to be found there.

Many Dutch cities have therefore been restored comprehensively in the recent past, and not just historic buildings, but entire streets and squares. Examples include the canals and *singels* that had long been filled in and have now been restored (see Chapter 5). The transport access corridors built in the 1960s and 1970s are now judged to have been a mistake, and the same is true for many buildings of the period. For better or worse, errors made earlier are being rectified.

An example is the Prins Bernhard viaduct. It was intended when it was built in 1970 to form part of an inner city ring for The Hague, until the planning of the route faltered at the Amsterdamse Veerkade. Thus, in the 1980s, a decision was taken to shorten the newly constructed Prins Bernhard viaduct again. Unfortunately, this corner of the city still does not work well.

The same cannot be said for the new building to replace what was known as the Maupoleum in Amsterdam. This building arrived in the 1960s along with the construction of the Wibautstraat corridor and the Waterlooplein roundabout, but simply could not be made to blend in with the surroundings. The building was therefore demolished and replaced by a new building that is more in keeping with the street.

The intention of other alterations was purely to repair a familiar cityscape, but some of the buildings created in the process are not historically authentic. For instance, the Grote Markt in Groningen was given a historicising new building complex instead of the modernistic building by the architect P. Zandstra, and it is now the turn of the north side the square near the Martini Church. And new buildings resembling fortifications and containing a multi-storey car park and a retail complex appeared in Zwolle on the site of the former rampart. A recurring topic of discussion in every alteration for urban restoration is the extent to which the aim should be achieved with contemporary interpretations and architecture, or by retaining the form of the original building.

However, Dutch city centres are still not splendid examples of restoration. They might more accurately be described as an attractive mix of historic elegance and untidy, small scale with a dash of traffic: a kind of urban crudeness. They also lack the metropolitan ambience of large, full pavements and tall buildings.

Much effort in the recent past has been put into improving non-commercial public spaces, such as parks, squares and riverbanks. More of the riverbanks in Rotterdam and Amsterdam are gradually being opened to the public, which sets the scene for new urban forms of waterside leisure activities. Paris and Barcelona set an example for how to approach public space. Interest is growing for new projects, such as extended use, even after dark and in our climate (possibly inspired by Lyon's lighting plan and Copenhagen's pavement cafés with outdoor heating).

Figure 3.5 Old and new building in the city centre of Groningen.

Source: Evelien Brandes.

Figure 3.6 Dordrecht.

Source: Evelien Brandes.

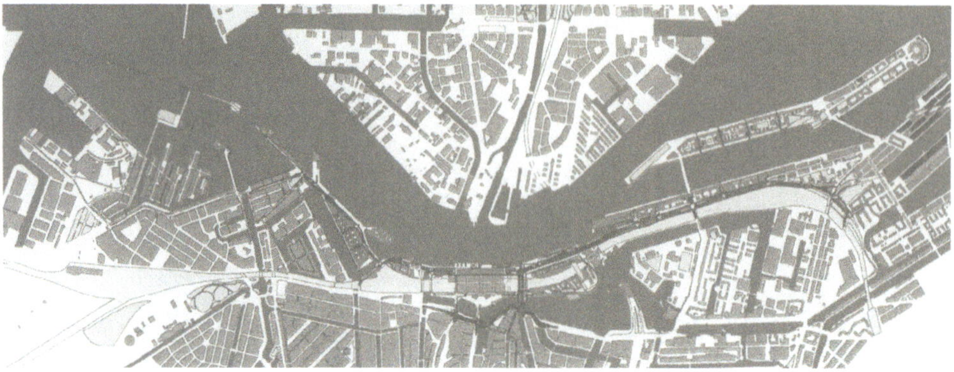

Figure 3.7 Plan for the IJ bank Amsterdam (see also color plate 7).

Source: Evelien Brandes.

With the exception of Rotterdam, the spatial structure of Dutch city centres was never fully compatible with the modern, large-scale centre functions of large retail concentrations, offices and high-rise buildings. However, these elements are inevitably attracted to the centre environment. Dordrecht is an example of where this potential conflict exists, with a combination of the small-scale city centre and its poor accessibility. The city centre itself consists of picturesque but narrow streets and enclosed harbours where traffic restrictions are in force. But the city centre is on the periphery of Dordrecht's urban area and is therefore accessible from only one side. The station is also some considerable distance away. All these factors may well make a visit to the city centre worthwhile, but hardly convenient. The city centre contains some major department stores, which are necessary to draw shoppers in to the zone. However, what these city centre businesses want is a large building and easier access by public transport and car. One of the important urban design issues for the city centre is to produce spatial plans that strike a proper balance between these conflicting factors.

3.3.2 Old industrial sites, harbour areas and other fault zones: new designated uses

The intention in the 1960s was to demolish the first expansion areas and the old urban districts, to create space for new urban developments around the centre. However, these plans met with formidable resistance. After many conflicts surrounding demolition and new building, new locations have been developed in the past thirty years that are designated for a combination of urban and residential uses.

Many cities have had, and still have, outdated industrial complexes and harbour sites with an ambience and possibly a view over a river that offer potential. Amsterdam's East Harbour Area and Rotterdam's Kop van Zuid contribute to the quality of the entire city. Some examples of current plans are the banks of the river IJ in Amsterdam (Houthavens, Westerdok Island, Oosterdok Island and Oostelijke Handelskade) and the Lloyd Pier in Rotterdam (Shipping and Transport College, studios and apartments).

Figure 3.8 Urban renewal.

Source: Evelien Brandes.

Other locations in the city fringes have also been taken into development for large-scale business activity, such as the Zuidas in Amsterdam, the Utrechtsebaan and surroundings in The Hague and Northeast Rotterdam. The design of these new intersections into multifunctional areas of merit is an important issue for the immediate future. Care must be taken in the development process not to allow central parts of the city to lose too much vitality and economic function. The point is to strike a balance that benefits the quality of the surroundings.

Amsterdam is a city in which the pre-war municipal districts described above have run into each other as a matter of course. Although these components are easily identifiable and distinct, the boundaries are not marked with physical barriers or intermediate transitional zones. Every expansion was designed with care: the encircling canal was kept intact and painstaking allowances made in the expansions. Berlage arranged for the streets in his expansions to link seamlessly with the then existing edge of the city. The integrity of structure that he created must be consolidated in the current issue. The aim is to absorb the current requirements, such as accessibility, heavier traffic and the constant need for more homes, within the existing structure, without making additions such as driving through new major access corridors.

Some other cities do have fault zones with space to offer. The issue there is to transform the poorly performing areas completely. For example, the gasworks and power station site is gradually disappearing from the centre of The Hague. This once isolated location, which was backed onto by the surrounding districts, now provides space for new activities and an opportunity to bind the existing districts. A large park, appropriately named 'De Verademing' (breath of fresh air), is a complete transformation that projects positively into the surrounding districts. The refurbished old buildings, new housing, combined residential and business properties and commerce are located at the edge of the park. The many rail lines in Utrecht form barriers in the urban fabric. Gradually, attempts have been made to create more and better links at strategic places, allowing the empty zones along the track to be built up more intensively. The area known as the Cartesius triangle between the rail lines to Amsterdam and The Hague is awaiting a new urban function. Fault zones of this kind are forming a new area in repairing the city.

3.3.3 The first and second rings: a renewal of the urban renewal?

The poorer districts from the period between 1880 and 1920 have been renovated drastically in recent decades. This was sometimes a matter of demolition and new building, producing a completely different urban structure, and sometimes only of replacing or renovating homes. The Old West and Crooswijk districts of Rotterdam, the Schilderswijk in The Hague, and the Dapperbuurt and the Kinkerbuurt in Amsterdam have all been dealt with rigorously. Although there can be no doubt that the quality of living and housing has improved, certain qualities have been lost. The number of shops and volume of business activity, the daylight in the streets and the aesthetics of the old building fronts have often deteriorated in these districts.

Figure 3.9 Garden town.

Source: Evelien Brandes.

A number of cities are investigating whether the output of the 1970s renewal, some good and some bad, should be tackled again. The areas concerned are the forgotten pieces, such as along the major approach roads and long retail ribbons. More attention is being given to the autonomous appeal of these old districts for first-time buyers and enterprising city people. The renewal of these districts does not have to rely exclusively on the government or housing associations. For instance, the Rivierenbuurt in The Hague had a very varied composition. The district was improved progressively by private residents together with a few limited new building and public space improvement exercises.

A glaring missed opportunity in the urban renewal period was that relatively low-density residential districts were built near stations, and moreover for a limited target group, who are not users of the station. This happened in, for example, the 'forgotten corner' near Utrechtsebaan and parts of the Station district in The Hague. Transformations such as an increase in building density, new frontages or entrances, taking the public space in hand, and demolition and new building; various strategies were investigated with a view to doing justice to the potentials of these locations.

The former urban renewal districts have a relatively large paved area, although they do sometimes have *singels* and canals. Opportunities for improving the water quality and the water system, possibly by increasing the water area, are addressed in the next chapter.

3.3.4 The residential areas of the interwar years: consolidation or demolition and new building?

The districts in the ring built in the interwar years are now about 75 years old. These residential areas are usually still satisfactory, because many homes were built with a degree of urban design vision and social inspiration. An important point is the composition of the building blocks, the relationship between the public space and the buildings, and the quiet surroundings with limited functional mix. Old South and Old West, the Berlage-designed Amsterdam expansions, have been restored completely and are now in a splendid condition. It was even decided belatedly to build a tower that originally had been overlooked in the restoration of Mercator Square in Old West. The architecture of 1930s homes has become a popular building style, which also sells well in imitation form in new expansion districts. Consolidation, also for individual occupation, was the main issue for many products of the interwar years.

The garden villages built in the 1930s were also often restored, despite the inadequate size of the homes and structural defects. Garden village Oostzaan in North Amsterdam was created in 1924 as an emergency housing complex and was a product of the vision of alderman Keppler and the architect Boeyinga. The controlling housing association wanted to replace the entire neighbourhood of 2800 dwellings in the renovation, but they underestimated the resistance from the residents. The housing, but in particular the urban green character, mattered a great deal to them. The conclusion of a lengthy consultation and planning period was therefore for restoring the district. Residents were required to pay far higher rents for the still small homes, which they thought was acceptable if their neighbourhood could be preserved.

On the other hand, not everything from this period is held in esteem. Amsterdam is not representative on all fronts for the current issues in Dutch cities. Both the original urban structure and the interest in living in the old city are generally less strong in cities like Rotterdam, The Hague and Utrecht than in Amsterdam. This has repercussions on various districts from the interwar years, which are usually further from the centre but lack the greenery and the expansion potential of the urban fringe. The housing quality and the internal urban quality are considered insufficient there to compensate for the disadvantages of the location.

Figure 3.10 Interwar area, ready for demolition.

Source: Evelien Brandes.

Figures 3.11.1, 3.11.2 Spangen, aerial view and renovation plan.

Source: Hulshof Architects Delft.

For example, large districts from the interwar years in The Hague, Transvaal, Duindorp and Spoorwijk, are now being demolished and rebuilt. Sweeping plans are being drawn up and executed in the Rotterdam districts of Spangen, Bospolder-Tussendijken and various residential areas in South Rotterdam. The choice between demolition and new building or consolidation is a hard one to make. Moreover, in Amsterdam itself, the entire urban area north of the river IJ, with large empty harbour sites and various garden districts from the interwar years, has issues that are totally different from the rest of the city. North Amsterdam is only now attracting interest, as a result of the North-South metro line construction. The major transformation and new housing issue for Amsterdam will be largely focused on the north in the coming period.

3.4 THE ISSUES IN THE PRE-WAR CITY BY THEME

3.4.1 Accessibility and the environment

Improving accessibility in the cities combined with reducing air pollution is one of the most important issues for the future. The most important spatial objectives for the cities are a healthy urban structure with sufficient greenery and water threading through, together with clean and space-intensive forms of mobility. The main points are the dynamics of the functions and the relationship with the residents. The issues described below can be considered as derivatives of the above.

3.4.2 Station areas

Station areas fulfill a crucial function for a city. Overall, the area around the station is the best accessible point from all parts of the region. Large flows of people are brought in through the station every day. Despite this hotspot quality, the surroundings of many stations are monotonous and poorly utilised. In the past, functions related to rail, such as sidings, shipping and post logistics, needed these locations for themselves. Many plans currently exist for developing precisely these areas more intensively. Under the heading of key projects, the national government is sharing responsibility for the transformation of the stations and their surroundings. A complete transformation is in progress in Amsterdam with plans for the banks of the river IJ, the Westerdok Island and the Oosterdok Island. In Rotterdam too, the area around the station is undergoing a metamorphosis. The arrival of the HSL needs not only new track but also a new station. The surroundings will benefit from this intersection with new routes and high-rise buildings. Planning has been ongoing for the Utrecht station area for many years. The area involved is the station, the Jaarbeurs exhibition centre, the Hoog Catharijne shopping centre, and the contact with the old city by putting water back in the Catharijnesingel.

Many station districts in large and medium-sized cities, such as Zwolle, Leiden, Breda, Amersfoort and Arnhem, had also either fallen into disuse or were occupied by sidings and second-rate business activity. These areas have recently been enhanced with new residential and office buildings, and further development is foreseen in these locations in the coming period. The objective is a mix of functions including housing, which provide for a permanent presence of people around the routes. The number of routes (and types of traffic) directly around the station demand multilayer solutions – underground, on ground level and elevated – in an attractive and, if at all possible, open and green setting.

3.4.3 Cyclists and pedestrians

A feature of Dutch cities is the use of bicycles, which contributes significantly to the accessibility and to a clean and healthy urban environment. What is needed in spatial terms is a good cycle infrastructure, consisting of safe cycle paths, through cycle lanes and sufficient parking facilities in the centres, near the stations and the housing. Many Dutch cities have been paying considerable attention to the subject in recent times, together with tackling the public space. Cycle facilities and attention to safety also enhance the livability in the city for children. Pavements and clear crossing places are vital in the city centres as well as in the residential districts, and they encourage more pedestrians to use the public space, which benefits the vitality. Trees lining the streets and green verges, where possible combined with (running) water, are perceived in densely built up and heavily used urban areas as linear buffers, which help counterbalance the road traffic. The issue in street face-lift plans is to integrate these elements into the limited space.

3.4.4 Infrastructure

New infrastructure plans are being executed in several Dutch cities to improve accessibility and the circulation of urban traffic. RandstadRail is being constructed in The Hague and Rotterdam, in some places in the form of new route sections, and in other places by altering existing tram and metro tracks. The influence of the project on the surroundings is enormous. Opportunities arise around the stops for creating new public space, for new functions and for increasing the building density. These projects also lead to renewal in an architectural sense. A spectacular feature is the elevated metro line along the Beatrixlaan in The Hague, which runs through a structure resembling a net stocking. The North-South underground metro line is under construction in Amsterdam. The streets will gain metro entrance buildings, Central Station and surroundings are to be upgraded, and the North Amsterdam urban district will be transformed structurally and the building density increased.

Express tram, metro and rail projects in the existing structure are generally drastic. While contributing connections, they are associated with damage to private and public interests in many respects. Arranging for the space around a project of this kind to heal properly is a complex urban design puzzle. Ideally, it is therefore formulated by the municipality as an urban design issue. These alterations therefore tend to be the subject of dispute.

There is discussion in Leiden on the express tram through the Breestraat in the city centre. The population of Leiden rejects this attack on the historic narrow street and is opting instead for a branch with the main line rail track and the stop in the existing station. The somewhat greater distance to the shops is seen as a price worth paying. The alternative that is to be executed has yet to be decided.

Utrecht's radial streets are being given a face-lift for the express bus lines to the Leidsche Rijn VINEX district to the west of the city. This will lead to a transformation of old continuous building strips into streets with new building walls, openings and bridges. Unfortunately, alterations of this kind can significantly disrupt existing qualities. The urban design department will endeavour to develop plans to compensate for this loss with items that can improve the performance and the townscape again.

3.4.5 Road traffic and car parks

Groningen, Haarlem, Delft and many small cities have succeeded in introducing traffic restrictions into the city centre. The understandable resistance from residents and shopkeepers notwithstanding, this has turned out to be a sustainable measure. On balance, the city centre has ultimately become a pleasant and economically sound environment. This is not a simple matter for larger cities with a greater city centre area. Traffic restrictions must at any rate be accompanied by a clear system of car parks around the city centre. The clever part is to integrate the car parks into locations a short distance from the centre. Suitable locations in the densely built-up area are often unavailable. Another problem is that underground parking is always extremely expensive in the Netherlands, because of building below sea level and the high groundwater table. Providing solutions for parking will continue to be a concern in the future.

Neither were the pre-war residential areas built with car ownership in mind. Much new parking space will have to be created, assuming estimates of a 50% expansion of the current fleet of cars (six million cars now and nine million in 2030). Measures of some kind will therefore be needed in many districts that are already full of parked cars. Some measures can be creative, but others may be difficult, with parking on courtyards, under or on the roofs of new buildings, in an old building, or in a ship tied up on the quay. New building complexes in the old city now usually have built-in parking spaces.

The increase in traffic in the cities is necessitating new construction or the expansion of existing roads. There are some major projects in development in this area. The Northwest

Figure 3.12 Parking problems in the second ring.

Source: Evelien Brandes.

Main Route, a system of existing roads in The Hague that are to be tunnelled and widened, is now in preparation. The motorway that runs through the middle of Maastricht, dividing the urban area into two, will be covered over in the coming period. It would appear that integrating large traffic flows does work better, despite the increase in traffic. On occasions, a city resolves to narrow an existing road, in conjunction with other works, as in the Phoenixstraat in Delft, which does not necessarily mean that the circulation has to suffer.

The increase in road traffic, combined with city buses that are almost never electrically propelled in the Netherlands, has a detrimental effect on city air quality. Yet other solutions will therefore have to be found in due course. Maybe cars and buses can be made more environmentally friendly, failing which the use and ownership of cars will have to be discouraged. If so, the effect will be to encourage the use of cycles and public transport. An increase in public transport can be achieved either through higher frequencies or expanding the network, possibly underground, which would have a significant impact on the spatial structure.

3.4.6 Green and water structure

Dutch cities in general are reasonably well equipped with green structures and open spaces (sometimes the river or a canal). However, the green structure in various cities has become shabby and the water polluted. Many cities have recently cleaned up their parks and water features, or are planning to do so. The removal of low vegetation and structures together with intensive maintenance of plants and shrubs has proved an effective way to stop the degeneration of most parks. The Zuiderpark in The Hague has been restored to its former glory in this way.

Opportunities for new parks with multiple potential uses are being sought in some large, fully built-up urban areas, such as Rotterdam's West and Feijenoord districts. The planners have their sights set on old railway junctions. Varkenoord Park in South Rotterdam has been created on a canopy over the sidings adjoining Bospolder-Tussendijken. Another success is Amsterdam's Westergas Park. The city is also attempting to provide new outdoor leisure activities, such as the large roof of the NEMO science centre on the river IJ, which gives the city centre resident from an apartment building with no balcony a chance to sit in the sun. IJburg beach is also in great demand on a summer's day. The advantage of these facilities is that they enable individual pastimes such as sunbathing and swimming in a collective and urban way.

An issue in several cities is to utilise the potentials of the existing parks again. They can be enriched among other things by housing around the edges, the laying of public gardens and the integration of opened sports grounds. Rotterdam's Zuiderpark and parts of the Kralingen woods are examples of eligible areas.

Particular attention must be given to the problems of the rivers, because the water level is rising and they have to discharge more water at unpredictable times. Where these rivers cross cities, the cities are faced with a choice: either to raise the dikes, or to widen the river bed. This water issue is addressed comprehensively in the next chapter.

3.4.7 Higher density of building and high-rise

The great demand for more homes in the limited area of the city centre is leading to meticulous searches for possible locations for increasing the density of housing. This is favourable for the city and urban life, because more residents mean more support for services. A high density of building is sometimes necessary to maintain population levels, because as the average number of house occupants decreases, the need for residential area increases.

Some cities are daring to introduce high-rise building. This style is very compatible with the urban structure of Rotterdam and Eindhoven. Rotterdam's new centre area and the open space along the river provide many suitable sites. The city council has been able to identify locations and expedite development for many years. For instance, the Hoge Heren (25 floors) appeared in North Rotterdam, and Montevideo, a fifty-floor block of flats, was built in South Rotterdam. The new housing culture ushered in by such developments is gradually taking root in the conservative Netherlands. Another high-rise apartment building was recently constructed in Tilburg city centre. Groningen, which has always experimented with high new buildings alongside old buildings, has situated apartment buildings near the historic city centre.

A more cautious approach to the cityscape is being taken in Amsterdam. The height throughout the old city is approximately four floors above ground level. New housing blocks in the middle of existing ones are therefore restricted to six floors. The only exception so far has been in the periphery along the A10, where taller buildings are allowed. The restrictions have not stopped Amsterdam increasing the building density within the ring. The sites of old bathhouses, schools, utilities, churches and warehouses are being converted for residential use. Other cities are also succeeding in similar meticulously detailed quests. Restoring old buildings and building new apartments are yielding new homes in small quantities in countless locations, for first-time buyers and older people moving back to the city.

3.4.8 Specific problems per city

Every city has a wide variety of issues in completing the plans, which have, perhaps necessarily, a detailed nature. A long period of aiming for homogeneity is now being followed by accentuating the distinct character of each city. The Hague is now giving much thought to the coastal zone and to the city's international function for peace and rights. Amsterdam is occupied with developing the banks of the river IJ, which are destined to become a glittering modern urban area, and the development of North Amsterdam is in the pipeline. Rotterdam has the great potential of the City Harbours, mainly in Southwest Rotterdam, for which flexible and phased plans are being developed. Moreover, medium-sized cities are also taking action: the university towns of Delft and Leiden want to link innovative business activity with their historic quality, which calls for new locations with a mix of housing and commercial premises. As was remarked in the introduction to this chapter, the existing urban area is changing constantly.

3.5 THE ISSUE IN THE PRE-WAR CITY

If municipal departments were asked to identify the urban design issue in the pre-war city for the coming period, they would give a wide variety of answers. The urban design departments are currently fully occupied with a wide range of projects and plans for many different sites in the urban area. The urban designer works 'remotely', as a director, because the spirit of the time is that 'the market has to make it'.

The large-scale approach is now in its final phase: the districts of the third ring from the interwar years are complete or in progress. Most are being restored, but some have been

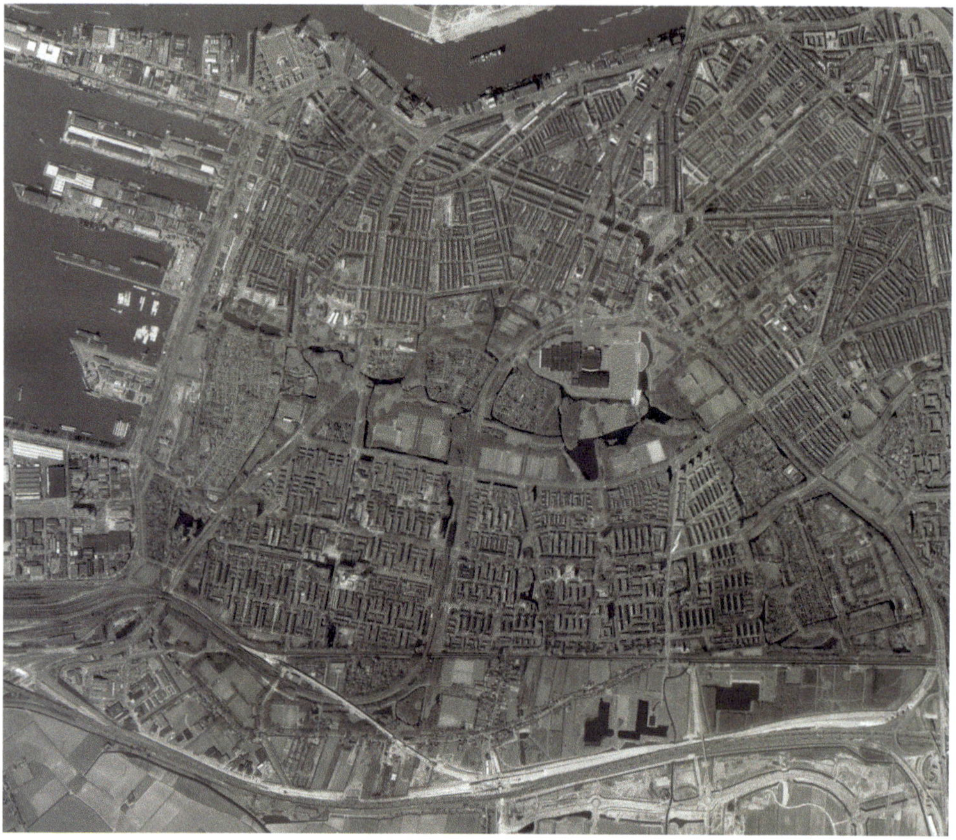

Figure 3.13 Aerial view of South Rotterdam (see also color plate 8).

Source: Google Earth.

or will be demolished wholesale and replaced by new building. The station districts around the HSL stations and in several medium-sized cities are being transformed drastically. There is a trend for restoring the historic parts of city centres to their former glory.

Besides completing the existing plans, the issue for the future of the pre-war city is to improve accessibility, the environment, greenery and water and to intensify the urban area. The focus will then be on consolidating the spatial quality for the residents and the urban functions. Furthermore, each city has various specific locations and topics, both large and small. The cities are developing into a differentiated, attractive and sustainable structure of urban, green, multifunctional and multicultural elements.

3.6 THE STATE OF AFFAIRS IN THE POST-WAR CITY

The restructuring of the post-war expansion is a current topic in almost every city in the Netherlands. The districts concerned are those built between 1945 and 1970. The Netherlands built two million homes in this period, which amounts to approximately 30% of the housing stock in 2000 (Hereijgers, Van Velzen 2001, p. 47). The restructuring tends to involve a broad range of viewpoints on tackling residential and other areas of this kind. An approach now commonly being opted for is one initiated by social or public housing, aimed at upgrading the housing stock. This renewal addresses the current demand of housing consumers (regarding type of housing and residential environment), with a focus on the demand for single-family dwellings.

The current land use and urban structure of the neighbourhoods, quarters or urban districts is not raised for discussion in many cases, but chosen as a starting point. However, from a spatial viewpoint, it is worthwhile contemplating whether the use of space meets current standards, or whether the urban structure of the urban district and urban regions might not offer new opportunities. It is necessary to investigate which design principles have proved sustainable and which could be called into question. The results of such investigations allow precise statements on which sites are valuable and which have scope for transformation. This insight will certainly be needed in order to use the opportunities presently offered by the post-war urban expansions. The water issue (which is discussed in the next chapter) plays a unique role in this respect.

3.7 THE POST-WAR URBAN EXPANSION

The post-war districts are based on the ideas of the modern movement and assume an 'open' city founded on light, air and space. These ideas were endorsed by the various

Figure 3.14 Masterplan Rotterdam.

Source: dS+V Rotterdam.

CIAM conferences. The CIAM philosophy was a reaction to the nineteenth century and early twentieth century urban infill development and urban expansion. The modern movement shifted the focus to the functional aspect of urban design, and in particular to separating the urban functions of housing, work, traffic, facilities and leisure activities.

Three design approaches can be identified in post-war urban expansion (Hereijgers, Van Velzen 2001, p. 49). The simplest is the allocation of lots along existing roads, which is common in village expansions. The second approach relies on pre-war expansion plans, such as the AUP in Amsterdam and the plan for the Southwest district of The Hague. The third and most widely applied approach is based on the 'neighbourhood concept' (Bos 1946), in which the design forms a complete whole. The starting point is the social structure of the district, as expressed in specific types of housing for each phase of life of the residents. This method was developed in Rotterdam for the Zuidwijk and Pendrecht districts, and was later copied by many other cities.

There was a substantial housing shortage shortly after the war, necessitating a high production of homes. This led to a rational approach to production (system building) with a focus on porch-accessed flats, generally four floors high with no lift. This evolved after the 1970s into blocks of flats of approximately ten floors with a lift, of which the Bijlmer is a striking example. The average housing occupancy at the time was about four people per dwelling, in a net housing density of fifty dwellings per hectare, and an average housing area of approximately seventy square metres. All these new building locations were situated on the fringes of existing cities.

The target group for the buildings was the working class, with an additional housing programme for elderly people and families without children.

3.8 THE COMMON APPROACH IN THE POST-WAR CITY

Social, economic and demographic changes have inevitably occurred in Dutch society in the fifty years or so that these housing estates have existed. The homes at the time were spacious and comfortable compared with those in the urban districts of the nineteenth century and the first half of the twentieth century. However, they are now far too small compared with the single-family homes built since 1980 in the centres of urban growth (from the Third Memorandum on Spatial Planning) and in the VINEX districts (from the Supplement to the Fourth Memorandum on Spatial Planning). Their comfort likewise no longer satisfies the housing standards of the time (such as central heating, a bathroom, garage, good insulation and a lift for apartments).

Upkeep was the starting point of the renewal of the nineteenth-century urban districts, by means of renovation to high standards. Demolition and new building was resorted to only in response to major structural defects. The principle applied in the renovation was affordable rents for the current residents: 'building for the neighbourhood'. The alterations were concerned with complying with the housing standards, such as having all storage areas on the ground floor and external spaces (balconies) located to receive daylight. Ground floor apartments were usually sacrificed in this process to make room for storage areas, and apartments were often combined to produce an acceptable housing area. Adding a floor (in the case of timber-frame construction) made it possible to limit the loss of dwellings to one-quarter. This loss was recovered by building in the centres of urban growth.

The approach outlined above is contradictory to the policy applied in the current restructuring areas, where building is encouraged in particular for the higher and middle-income groups. Another consideration is reducing the quantity of rented housing in favour of owner-occupied housing. There has been a shift in Parkstad (in Amsterdam's western garden towns) from 76% social housing to 45% in the new situation, in favour of 'expensive' rented (15%) and owner-occupied (40%) housing (Bureau Parkstad 2004, p. 34).

Besides the difference in policy, the restructuring areas have also seen demolition on a large-scale. For instance, one-quarter of the stock was demolished in Parkstad. The proportion in some of Rotterdam's restructuring districts was approximately one-third.

The problem with the existing stock of the post-war districts is again the small housing area (on average 70 square metres), rather than the structural state of the dwellings. There is now also considerable demand for single-family homes with their own garden, while much of the stock in the post-war urban expansion consists of multi-family dwellings, usually porch-accessed flats with no lift.

The policy in Rotterdam, for example, is for 60% single-family dwellings among the newly built homes in the restructuring areas. However, the ratio of single-family dwellings to multi-family dwellings is one in four, which means that many porch-accessed flats and blocks of flats have to be demolished in order to satisfy the demand for single-family dwellings, in particular for the middle-income group. In addition, the space taken up by single-family dwellings with their own garage is totally different from that of the porch-accessed flats.

Finally, the communal greenery surrounding these homes must be converted into the private greenery of the single-family dwellings (i.e. gardens). But this is not all: the necessary land area per dwelling is many times greater than for multi-storey development. The issue for urban developers is to devise appropriate solutions, which can possibly be solved not only on the level of the residential complexes themselves, but more in particular on the level of the district and the scale level of the city.

3.9 NEW OPPORTUNITIES IN RESTRUCTURING THE POST-WAR CITY

The following three relevant levels for releasing the potential of restructuring areas are identified in the book *De naoorlogse stad. Een hedendaagse ontwerpopgave* (*The post-war city. A contemporary design issue*)(Hereijgers, Van Velzen 2001, pp. 50 and 111):

- the strategic level, the location in the urban field of today;
- the level of the urban structure;
- the project level, in other words the firm building plans.

Involving the first two levels above in the design issue creates additional opportunities for giving meaning to the renewal for the entire city and the associated region.

The strategic level addresses the location of the restructuring areas in the city of today, because even if these areas were at the edge of the city at the time of the original development, they are now usually at the centre of the urban network. They are therefore easily accessible by car and public transport, but they are also within cycling distance of the city centre and from the regional green structure.

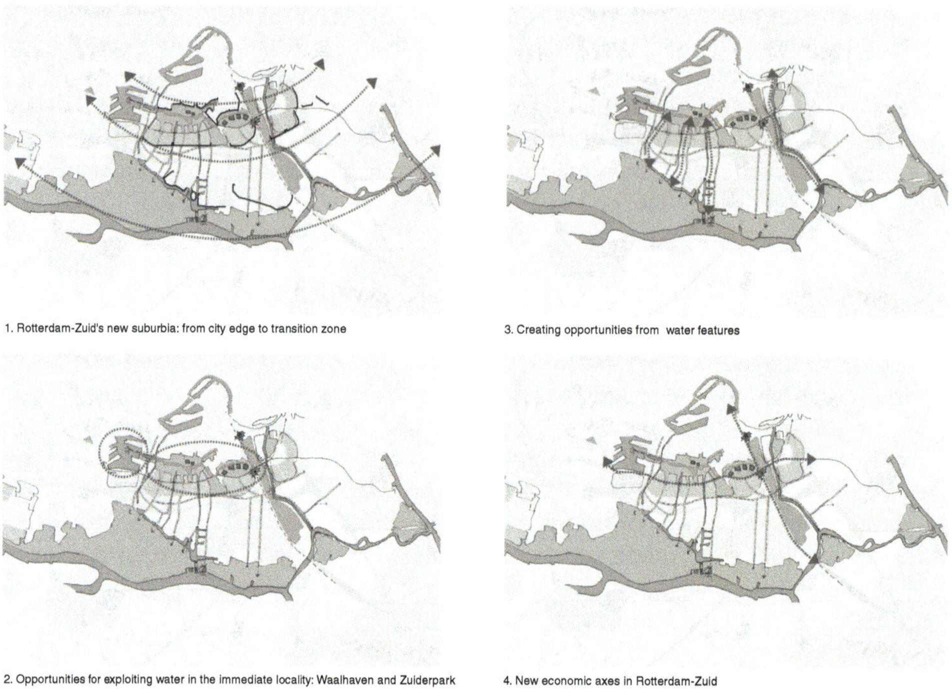

Figure 3.15 New positions for Rotterdam (see also color plate 9).

Source: dS+V Rotterdam.

Where the infrastructure exists, for example at a public transport stop, the restructuring may use new opportunities, by means of a specific programme around the stop (higher densities, special housing for young people and elderly people, care and education facilities and other economic functions). It is possible for a programme to accommodate an existing or planned green 'artery' (in combination with water), ultimately creating various residential environments that do justice to the location in the urban network.

Repositioning the restructuring districts on a strategic level has direct consequences for the urban structure and might lead to redesigning the infrastructure and the green and water structures, as well as to identifying new residential areas or locations for facilities.

An obvious implication is that the desired urban programming is oriented to this modified spatial structure.

All in all, this leads to a totally different approach to the restructuring areas than the one now commonly applied, in which only the housing stock is upgraded, and which is embedded in area agreements made between the municipalities and the housing associations in mid-2005 for 56 restructuring areas prioritised by the Ministry of Housing, Spatial Planning and the Environment. The area agreements provide for an additional government financial contribution for these areas between 2005 and 2015 (Urban Renewal Investment Budget, ISV, second tranche).

3.9.1 Establishing the profile of Rotterdam's southern garden towns

As an illustration and in conclusion, the profile established for Rotterdam's southern garden towns (Rotterdam Urban Design and Public Housing Department 2006) is discussed below. The southern garden towns consist of the four housing estates Zuidwijk, Pendrecht, Lombardijen and IJsselmonde. The joint housing stock of these housing estates is 30,000 dwellings, 5000 of which are single-family dwellings (Zweerink 2005, pp. 266–267).

Area agreements were also made between the municipality and the housing corporations for the southern garden towns, regarding the homes to be built and the associated outdoor area between 2005 and 2015. The scope is 4500 newly built dwellings, including 60% apartments and 40% single-family dwellings, and 80 hectares of redesigned outdoor area. The water issue is not included (Urban Design and Public Housing Department, Rotterdam City Development Corporation 2005).

The Urban Design and Public Housing Department initiated an investigation in early 2006 into the potential of these housing estates on both the strategic and urban structure levels. The results of this investigation are endorsed by the municipality and the housing associations involved in the 'Pact for South', which has led to a drastic reconsideration of the area agreements.

The central issue of this investigation was the aim expressed by politicians of making Rotterdam an attractive residential city, for everyone and for the middle class in particular. This group has left Rotterdam in large numbers in recent decades for suburban residential areas (VINEX districts), usually as a deliberate choice, but sometimes because of the lack of any alternative within the (post-war) city.

Stadsmensen. Levenswijze en woonambities van stedelijke middengroepen (*City people. The way of life and housing ambitions of urban middle classes*) divides these middle classes into three groups: the social climbers, the urban families and the old Rotterdam residents (Karsten, Reijndorp and Van der Zwaard 2006). Each subgroup naturally has its own set of housing wishes (its housing profile). The area agreements that were made paid insufficient attention to these three groups. Naturally, the current residents who want to stay are also included.

The starting point for the urban renewal of the southern garden towns is to make and keep them attractive for these target groups. The implication for the programme is that 60% of the dwellings are to be realised as single-family dwellings and 40% as apartments. A comparison with the existing stock, less than 20% of which is single-family dwellings, shows that very many multi-storey buildings will have to be demolished. Regarding the area agreements that were entered into, this means a shift in the programme from 40% to 60% newly built single-family dwellings. The extent of the loss of housing caused by this shift has been investigated (Breebaart and Huffstadt 2006). The models developed for this purpose show that this aspect is not necessarily relevant, provided the design is to a high standard.

The research report *Profilering Zuidelijke Tuinsteden* (*Establishing the profile of the southern garden towns*) builds on three earlier spatial studies, which, rather than starting from the project level alone, also considered the strategic level and the urban design level of the district. The cultural and historical values of the southern garden towns are among the subjects covered in the studies *Van Ommoord tot Pendrecht*

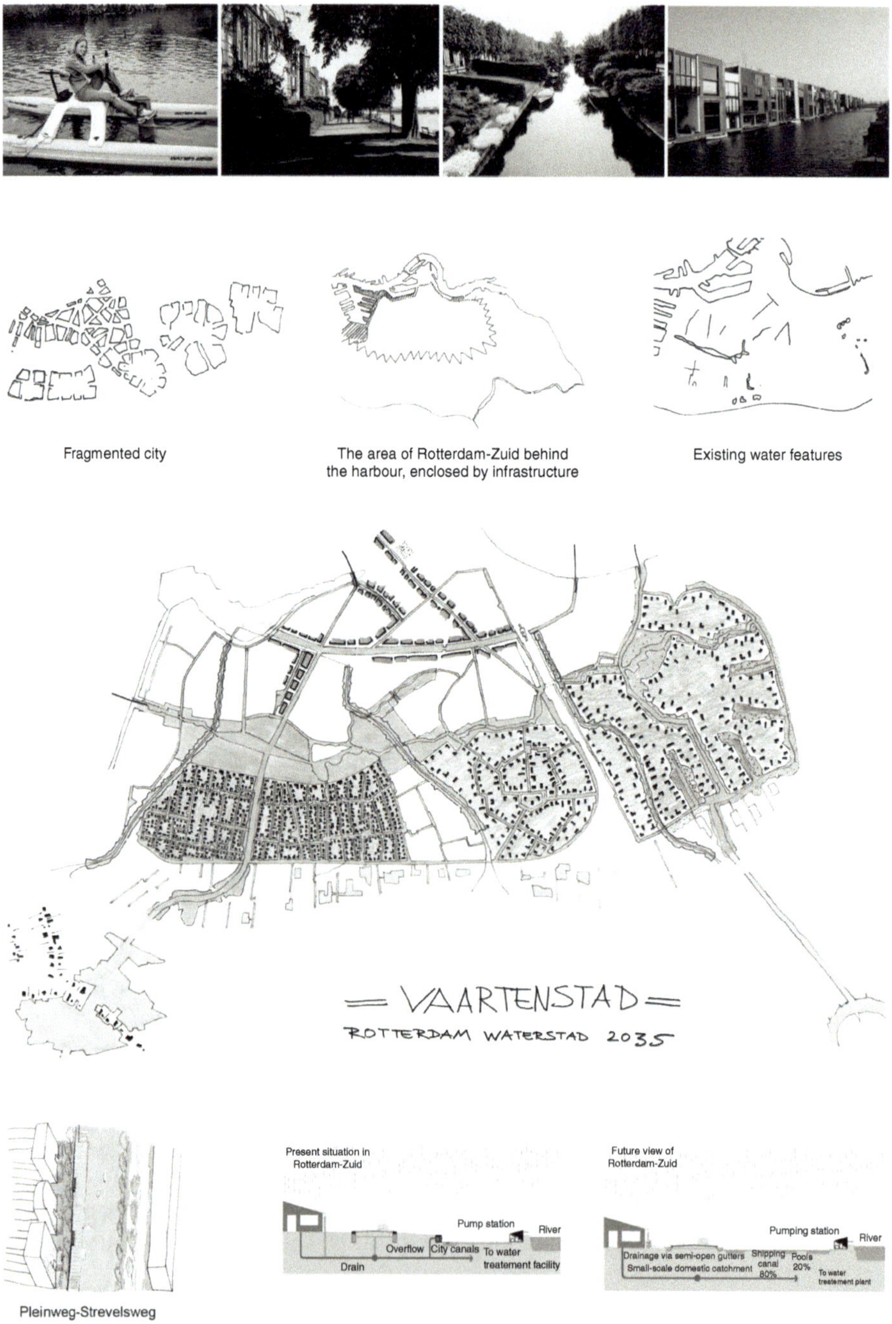

Figure 3.16 Water map: South Rotterdam canal city (see also color plate 10).

Source: dS+V Rotterdam.

(*From Ommoord to Pendrecht*) (Zweerink 2005), *Toekomstvisie Rotterdam Zuid 2030* (*Vision for the future of South Rotterdam 2030*) (Urban Design and Public Housing Department, Rotterdam City Development Corporation, Kaap3, De Beuk and Van Gilst 1999) and *Rotterdam Waterstad 2035* (*Rotterdam Water City 2035*) (De Greef and Urban Design and Public Housing Department 2005). *Rotterdam Water City 2035* approaches the water issue for South Rotterdam as not only a technical problem, but more in particular as an integral component of the approach on both strategic and district structure levels (De Greef *et al.* 2005). The new *Vaartenstad* was devised as a leading concept for South Rotterdam, consisting of water links running north-south. It is necessary to create approximately 45 hectares of additional area for South Rotterdam.

The report *Profilering Zuidelijke Tuinsteden* (*Establishing the profile of the southern garden towns*) uses these studies to arrive at a redesign of both the green and water structures, as well as the spatial-economic structure. It describes an east-west green and water structure centred on the Zuiderpark (known as the '*Gordel van Smaragd*') in combination with new north-south links with the current urban fringe (Regiopark and the Oude Maas river).

New reference points for new unique residential environments will be created by combining these newly designed structures with the water issue and the housing issue. Two economic axes can also be identified. One axis is a north-south oriented zone parallel to the Rotterdam-Dordrecht rail line (*Stedenbaan*), and the other is an east-west axis linking all districts together.

Figure 3.17 Ambitions South Rotterdam (see also color plate 11).

Source: dS+V Rotterdam.

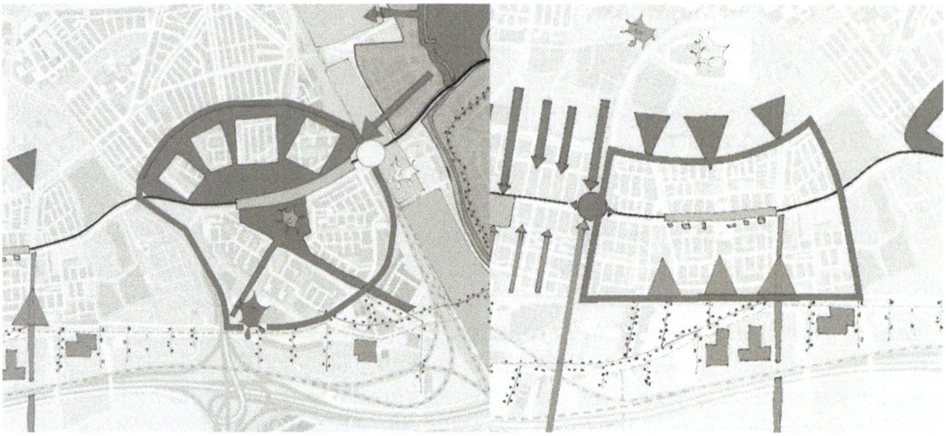

Figures 3.18, 3.19 Profiles Lombardijen and Zuidwijk (see also color plates 12,13).

Source: dS+V Rotterdam.

This approach produces a new spatial frame for all the southern garden towns, which serves as a framework for the further renewal on the district and project levels (the building issues). What this means specifically is that a profile is drawn up for each district as a development perspective for defining a programme for each project.

Dealing with restructuring the post-war urban expansion in this way not only addresses the problems of the existing stock, but also offers a perspective for new spatial developments that do justice to the strategic location.

Furthermore, space is also created for building additional housing and other programmes, schools, care institutions and businesses. The water issue too can be solved, not only as a technical problem, but as a spatial vehicle for new residential environments.

The best use is thus made of the unique location and culturally and historically valuable structure for making the post-war urban expansions attractive again, not only for the current residents, but also for people who were obliged to leave the city and are now keen to return. Fortunately, it can be observed that this direction is also been opted for in Amsterdam (Parkstad and the renewals of the New West district and the Bijlmer) and The Hague (Southwest).

Chapter 4

The water issues in the existing city

Sybrand TJALLINGII

Faculty of Architecture, Delft University of Technology, Delft, The Netherlands

4.1 INTRODUCTION

Old city centres, urban expansions from the interwar years and post-war districts are part of the existing city, which must constantly adapt to economic, social and ecological developments. Water issues are currently an important aspect of these transformation processes. This chapter explains the component parts of these issues and how the role of water can be integrated into urban renewal processes. Each situation is different, and demands its own approach and its own design. Designers and planners therefore have an important task to fulfil. They have to deliver custom work, which is not to say that everything is different each time. It is not necessary to keep reinventing the wheel, but designers can and must allow themselves to be guided in each plan by a few general

Figure 4.1 Willemstad.

Source: Fransje Hooimeijer.

principles derived from the lessons of the past and the requirements for the future. This chapter is about those principles. A proper understanding of the water issues first requires some insight into the water flows themselves: the sea, the rivers, *boezem* (out- and inlet waterway) and polder waters, rainwater, groundwater, drinking water and wastewater.

The 'water issue' tends in practice to be presented rather simplistically in statistics. For instance, the *Fifth National Policy Document on Spatial Planning* (2001) forecasts that 375,000 hectares will be needed for water storage by 2030. Some district water boards insist that 10% of the space in new urban areas consist of water. The message is clear: water takes space, much space. It would appear obvious, yet it is an oversimplification. Water storage is not about space, but about volume: not about square metres, but cubic metres. Water storage moreover is not the only issue. Water quality is important too. Water design is concerned with a water system that solves the problems of too little, too much and too dirty. Furthermore, designing is about more than solving problems alone. Water is part of the quality of roofs, streets, squares and parks. The design issue in these urban spaces is about creating conditions for ecological, social and economic quality. Interesting functional combinations emerge if we handle water systems intelligently in designing urban areas, and it is even possible to argue that water takes up no space.

4.2 WATER FLOWS

4.2.1 The sea: from resistance to resilience

The sea level is set to rise significantly in the coming century, while the ground level in the coastal provinces is falling. Almost nine million people live in the parts of the Netherlands affected, and 60% of our gross national product is earned there. Coastal protection is the most important water issue in towns, such as Flushing and Scheveningen, located right by the sea. The *Third Policy Document on the Coastal Areas* (2000) and the *National Spatial Strategy* (2004) show that there has been a reversal in the thinking surrounding coastal protection in recent years: there is now less opposition to natural forces and more going along with them. The natural coastal foundation of the Netherlands consists of sand: from the shallow sea up to and including the dunes. What going along with nature mainly comes down to is opting for the natural sand movement as the organising principle. Dune erosion repairs itself naturally as sandbanks regenerate, drift, and form new dunes. Artificial sand replenishment is carried out where the dynamic equilibrium has been disturbed, either on the beach or under water.

The key word in this strategy is *resilience*. The term *dynamic maintenance* of the coast is also used. Hard constructions are static and not resilient, because they limit the sand movement, and extremely expensive measures may be needed to maintain them. The building policy that this implies for coastal towns and villages is one of 'yes, provided that' within existing building areas and 'no, unless' outside. It is conceivable to combine urban development with reinforcing the coast. For example, plans have been made in the past thirty years for one of the most vulnerable parts of the Westland region near Monster, towards both the sea and the land. The urgency of coastal protection is increasing all the time with the expected climate change, but the conflicts of interest between safety, housing and glasshouse cultivation thus far have prevented agreement on the spatial and financial possibilities. The design issue here is not primarily to develop a spatial vision, but to bring the multiple stakeholders into line, with a view to guaranteeing the objectives of coastal safety.

4.2.2 The rivers: from raising dikes to space for the river

River levels will fluctuate more. The peaks will be higher, and the troughs deeper because of changing precipitation patterns, exacerbated by drainage and the increasing paved area in upstream cities. This development leads to several issues for the Dutch river towns. For example, Nijmegen, with its city centre rising high above the river Waal, is under less pressure than a city like Deventer, with its riverfront at water level. Even more intrusive are the issues for a city like Dordrecht, where the old city and its harbours were built outside the dike and the flood defence runs through the middle of the city. The central issue here is defence against flooding, but this is merely treating the symptoms from the viewpoint of river management.

Here too, there has been a reversal in thinking: from raising dikes to *Space for the Rivers* (1997–2000), the policy in which the Ministry of Transport, Public Works and

Figure 4.2 The coast of South-Holland (see also color plate 14).

Source: Province South-Holland.

Figure 4.3 The river Rhine at Tolkamer.

Source: Sybrand Tjallingii.

Water Management and the Ministry of Housing, Spatial Planning and the Environment state that the central river management issue is improving the discharge capacity through spatial interventions. One implication of the above is the construction of secondary channels and of green or new rivers as bypasses for the bottlenecks in floodwater drainage. The combination of urbanisation and new rivers is interesting in that it offers win-win opportunities. Living on the water is popular. This idea has prompted plans for living outside the dike, some of which have been implemented.

4.2.3 Streams: from straightening to 'remeandering'

Higher precipitation peaks and longer periods of drought also have a direct impact on water management in the higher parts of the Netherlands, where streams are responsible for drainage. For example, the towns in Twente and Brabant have been confronted with flooding for some considerable time, and climate change is making matters worse. The causes here are the improved drainage of agricultural land in combination with the channelling of streams and the increased paved area attributable to urbanisation. The result is discharge peaks, which cause problems mainly downstream.

The reversal in thinking on streams is from straightening to 'remeandering'. The strategy in the cities has changed from rapid discharge to one of first hold, then store and only then discharge. This is the three-step strategy of the 21st Century Water

Figure 4.4 *Boezem* and polder waters: integrated in Ypenburg.

Source: Sybrand Tjallingii.

Management Committee. The water issue of 'hold and store' is often possible by means of infiltration in the stream landscape. It is achieved in the urban design by constructing ground level or underground infiltration devices. An example is the swale or *wadi*, which is a green trench that fills up in a rainstorm and then allows the water to infiltrate into the groundwater. Swales or *wadis* in a residential area are part of the green structure, and with careful design they occupy no additional space. This kind of water storage is now already standard practice in many municipalities. This solution is sometimes also possible outside the stream landscape, as in parts of Utrecht's Leidsche Rijn VINEX district, to facilitate the optimum use of subsurface storage. The design issue here is proper integration of infiltration facilities, swales and stream valleys in the town. Good examples have been implemented in Hengelo, Enschede and elsewhere.

There is also a qualitative water issue in the urban areas of the stream landscape, which is related to the often-heavy effluent burden on streams from water treatment plants (the treated wastewater discharged into surface water), overflow water from mixed sewer systems, and fertiliser and pesticides leached from farmland. We observe in cities with an effective sewer policy and well controlled industrial discharges that the dirtier water no longer comes from the city but from the rural area. The issue for urban water policy is to prevent clean urban water mixing with polluted agricultural water. An example of what this can mean is the Zaartpark in the city of Breda, which was designed to allow the clean urban water to benefit nature conservation and leisure activities.

4.2.4 *Boezem* and polder waters: from rapid discharge to hold and store

The water issue in the polder landscape is linked directly with the rainwater flow. Precipitation peaks lead firstly to nuisance in the polders, which lie 'upstream' in the drainage sequence. The *boezem* waters have a fixed maximum level, and drainage stops if heavy rainfall causes the level to be exceeded, so that the rainwater collects in the polder. Most towns and cities in the west and north of the Netherlands encounter this situation. Delfland is an example of a region in which the situation has become urgent following a trebling of the paved area in half a century, because of both urbanisation and glasshouse horticulture.

The issue was long formulated as one of improving drainage and enhancing the pumping capacity. The issue in the new approach is one of creating *peak storage*, which may suggest increasing the area of the *boezems*. However, this measure makes only a small contribution to solving the problem of the entire *boezem* zone. It is more effective to look first on polder water level for opportunities for holding and storing the water. The younger polder town districts often have more green space that can be adapted for a role in rainwater storage. The impact on the district itself is considerable, while it makes it easier to link the water issue with the urban design issue.

From the *boezem*, water is let into the polders and the towns in them during periods of drought each summer. Unfortunately, the water is of inferior quality. Solving a quantity problem thus creates a quality problem. Storage on polder water level in this case presents an opportunity to retain water in the wet winter season to last through the dry summer. This *seasonal storage*, when combined with peak storage, gives surface water a double water management role. Furthermore, surface water in particular offers opportunities for attractive residential environments. An essential aspect is how the level fluctuates. A fixed level was traditionally enforced in the open waters, but that provides no storage. The water can play a part in storage only if it is allowed to rise temporarily. This however has consequences for the design of all the banks. Allowances must be made in the living on the water combination for boat owners to navigate the *boezem* waters, and thus extend the range of trips they are able to make. Bridges need to be higher than required (which is also beneficial for skating). If a boat is moored in a polder lake, it will always have to negotiate a lock when moving further afield. The design issue for peak and seasonal storage often resides in the urban fringe, where more space is available for water and for level fluctuations.

4.2.5 Rainwater: from discharge to hold and store

Precipitation peaks and troughs were discussed above as they relate to the water flow of rivers, streams, and *boezem* and polder waters. If we wish to tackle the problems at source, it is important to establish what can possibly be done with rain between the time it falls from the sky and the time it arrives in the surface water. Rainwater falls on roofs, buildings, squares, gardens and parks, and it is clear that there will be fewer peak problems 'downstream' if the rainwater flow is held or used in these places. But there are other possibilities. For example 'green roofs' retard the rainwater flow, which is why many German cities prescribe them for flat roofs. Green roofs and roof gardens have a wider significance, in that they protect the roof construction, look attractive, and evaporation from them helps cool the hot city climate. However, fertilising these roofs causes eutrophication.

Figure 4.5 Storm water catchments in Den Bosch.

Source: Sybrand Tjallingii.

Figure 4.6 Ground water: infiltration at the Waalsprong in Nijmegen.

Source: Sybrand Tjallingii.

The design studies for the Utrecht Centre Project involved an exercise in which the large building complexes were viewed as a funnel for collecting rainwater instead of as an umbrella from which the water runs away (Claringbould and Hinterleitner 2005). The water collected is used later for inner gardens, which also play a part in cooling large glazed spaces. Other obvious uses of rainwater are in flushing toilets and for cleaning purposes, which would save drinking water.

The water issue in this case is therefore to use water in and around buildings. The resulting design issue calls for creative architects who use water to make buildings better and more attractive. This is a way for renewal projects in old city centres with little surface water, or where the surface water has a fixed level, to contribute substantially to the water issues.

4.2.6 Groundwater: from pumping to controlling

The groundwater issue is not linked directly with climate change but is associated with the extraction of drinking water and water for industry, and with the drainage of agricultural and residential areas and industrial sites.

Dewatering and drainage cause subsidence in the wet peat soils of the west Netherlands, leading in turn to high groundwater levels, the response to which is to lower the polder water level. A consequence is further subsidence, and a downward spiral. The ground in large parts of Holland has subsided by approximately five metres in the past thousand years, and is continuing to do so. Public authorities and private landowners in cities that are affected, such as Gouda and Rotterdam, are obliged to pay large sums for raising gardens, parks and streets. Attempts are made in preparing sites for building in new districts to create a favourable initial situation by creating a groundwater level of approximately one metre below ground level. The issue is sometimes formulated as *groundwater neutral building*. This has no direct impact on buildings standing firmly on foundation piles. The issue is in the public space and in the gardens. Experiments are being carried out with floating constructions in sensitive areas, such as roads on expanded polystyrene. The issue demands the development of joint strategies on the part of designers and water managers.

There are other problems on sandy ground, where the groundwater level can be lowered by extraction for drinking water and industry, the deep drainage of farmland and sometimes also by a sharp increase in paved area, which impedes rainwater infiltration into the groundwater. The consequence is falling water tables, which in turn may prompt the installation of irrigation systems, thus lowering water levels even further. In housing areas, falling water tables are less problematic, but in agriculture, the situation is different. Also, the managers of nature reserves are not happy with streams falling dry, as happens along the foothills of the Veluwe in the central part of the Netherlands. Special problems occur in cities such as Enschede, Eindhoven and Delft, where industry pumped up large quantities of groundwater in the past. The pumping stopped together with the industry, and the resulting upward groundwater level trend is causing problems in crawl spaces, deep basements and underground car parks. The same problem is also occurring in urban areas along the inner dune fringe now that drinking water extraction from shallow groundwater has stopped.

The issue for the long term is to manage the stock rather than consume it. One of the options is to discontinue groundwater pumping and, instead, let the water first naturally come to surface in wetlands near the foot of the hills. These wetlands could then serve as a source of drinking water production for households and the industry. The Amsterdam drinking water production from the lakes along the river Vecht, at the foot of the Utrecht Hillridge, demonstrates the feasibility of this option. A genuine solution demands cooperation between spatial designers and water managers on a regional level. On the district design level, the issue is groundwater neutral building. This has led, for example, to a design for a 'leaking city' in Tilburg, the green design of which includes infiltration facilities for topping up the groundwater with rainwater, compensating the loss caused by the increase of the paved surface.

4.2.7 Drinking water: from wastage to careful use

The drinking water flow in the higher regions of the Netherlands usually starts with pumping up groundwater. The problems and the alternative solutions in this situation were discussed above under groundwater flow. The drinking water in lower parts of the Netherlands is usually extracted from surface water, which may be river water. For example, water from the river Maas ends up in Rotterdam via storage reservoirs in the Biesbosch. Traditionally, Amsterdam had two networks: 'Vecht Left, Dune Right', was inscribed on the old manhole covers in the streets. The Vecht water comes from the Vecht lakes, as described above. The Hague also drinks river water via the dunes. There are fewer problems associated with the use of surface water than with extraction from groundwater. The main reason for preferring groundwater extraction is to avoid the higher treatment costs.

The issue for managing the drinking water flow after extraction is oriented to preventing waste. Flushing toilets, cleaning, watering the garden and washing the car do not require water that is good enough to drink. Several experiments have been conducted

Figure 4.7 Drinking water: the Biesbosch.

Source: Sybrand Tjallingii.

Figure 4.8 Wastewater: Voorburg.

Source: Sybrand Tjallingii.

in the past ten years involving the delivery of two qualities of water, drinking water and (less purified) utility water, using two networks. Problems occurred in the practical experiments. Designing two networks presents no problems, but the implementation did not always run smoothly. Reversed connections, sometimes even by the water company's own engineers, ultimately led to the experiments being abandoned. The issue in this case is therefore mainly one of organising management with care and careful use. In large buildings, the organisation may be less of a problem, because an internal technical services department will often be involved. A decision not to use drinking water for flushing toilets and (sometimes) for washing machine water is also possible for homes. To avoid dual networks at the district level, the decision to use rainwater for flushing toilets and cleaning is also possible at the home design level.

4.2.8 Wastewater: from making clean to keeping clean

Until the 1950s and 1960s, canals in many cities were being filled in, partly because the stinking wastewater was a source of irritation. Fortunately, the time of black, stinking canals has passed. The successful application of the Pollution of Surface Waters Act (1969) is the main reason that water quality has improved considerably. The entire urban area is connected to water treatment plants and industrial discharges are cleaned. However, water quality problems have not yet disappeared from the city. For instance, *eutrophication* is a general problem: surface water acquires too high a nutrient content because of the use of fertilisers in gardens and parks and from waste, leading to algal bloom and, sometimes, dead fish, especially in shallow and easily heated ditches and urban ponds. Paradoxically, the water treatment plants are also a source of eutrophication, in particular because the phosphate content of the effluent after treatment is still too high. An additional phosphate trap is now being built into most plants to alleviate this problem.

A problem area closely connected with redesign issues is that of the combined sewers. Most older urban districts have a combined sewer system that connects both wastewater and rainwater and discharges it to the wastewater plant. This practice is detrimental to the treatment process, because the irregular arrival of large quantities of cleaner rainwater disturbs the purification process. Other problems occur in the city. The storage and discharge capacity of the combined sewers is inadequate in heavy storms, leading to *sewer overflow*, causing diluted wastewater mixed with sewer sludge to end up in the canals. This is currently the most important source of pollution in urban waters. The main solution to these problems until recently was to collect the overflow water temporarily in storage and settling basins. There has been a turnaround in thinking on this subject too, with the emphasis shifting to preventing problems by *disconnecting* the paved area, so that storm water no longer runs into the sewers.

Disconnecting can be achieved by constructing a new network of 'storm drains', but another way of dealing with storm water is to lead it through open gutters to swales, ponds or canals, or using it in buildings, as discussed above under the rainwater flow. The open gutter approach is a true design issue, which can be considerably less expensive than constructing an additional storm drain network. In some cases, such as busy roads, market places and bus stations, the quality of run-off water can be a problem. In these cases supplementary source control measures are required to prevent surface water pollution. These measures may include technical filters and filtering vegetation zones along banks. Creative design can find solutions that fit the situation. Many projects demonstrate the creative integration of rainwater in open space design. Yet, the water boards tend to be reluctant to embrace the open gutter solution to the disconnecting rainwater issue. The main reason is the complexity of open space and the many parties involved. For the technology-based organisation of the water boards, it is a lot easier to calculate and budget expensive storm drains than to engage in an interactive process with many parties.

The five flows each have their own issues, problems and opportunities. Yet, the role of the rain seems crucial. Both, the contribution of urban planners to sustainable water management and the contribution of water to the quality of life in cities are closely related to the role of the rain in urban design. The strategies for multiple water flows converge here: rainwater storage prevents floods and the fall of groundwater tables; rainwater collection may reduce drinking water consumption; disconnecting rainwater from the combined sewers and run-off quality control will improve surface water quality.

Evidently, these strategies do not only apply to the polder landscapes and the higher parts of the Netherlands. They have a more general meaning.

4.3 GUIDING PRINCIPLES: COMPREHENSIVE AND SUSTAINABLE

An analysis of the water flows has led to a keener formulation of both the water issues and the directly related design issues. These issues are summarised and elaborated below into guiding principles. The principles give direction to designers' creative quest, but are not standard solutions. Neither are they rigid dogmas, although they are formulated as firmly as possible. It is always possible in practice to apply them more flexibly.

The purpose of the guiding principles is to help find *integrated* and *sustainable* solutions. One-sided sector plans drawn up with the sights set on the short term are too common in practice. The flow analyses have shown that these plans often become part of the problem. Integrated and sustainable are fine words, but what do they mean in this context?

- *Integrated* is related to coming together. The first aspect is the connecting of the water flows discussed previously. It is also important within the framework of urban development to have cohesion between water policy, urban design and other policy.
- *Sustainable* adds that water design and management serve not only to improve the quality of life in the current city, but also have to offer prospects for the future values, related to use, perception and nature.

We can now view the issues for the various water flows as further refined principles for water flow management. In the context of this book – the role of water in renewing the existing city – we can reduce them to two core principles (the first two principles). Coastal protection is not considered here, because the urban areas involved in the Netherlands are few in number and differ considerably, and because there is little experience so far with the link between dynamic coastline maintenance and urban development.

An integrated approach addresses the meaning not only of urban areas for water management, but also, conversely, of water for the quality of urban *areas*. This leads to three new principles based on experience with sustainable urban development and sustainable water management (numbers 3, 4 and 5) (Tjallingii 1996; Kwaadsteniet, Jonkhof and Tjallingii 2000).

Finally, there are also principles for the planning process that are vital to a sensible role for water in the city that is supported by all *actors*, whether residents, users, or managers. The two last principles are based mainly on *Water in Drievoud* (*Water in Triplicate*), which discusses the results of an investigation into various approaches to urban water planning (Tjallingii and Van de Ven 2005).

4.3.1 Holding rainwater and keeping it clean

The first principle encompasses all the issues for water flows, from *peak storage* to *disconnection*. It highlights the central role of rainwater in designing with water.

4.3.2 Giving space to river discharge

A relevant problem in the river area is flooding, which is on a completely different scale and demands a guiding principle of its own.

4.3.3 Coordinating water use and water management and making them visible in the plan

This means that water use must be attuned to the manageable capacity, so that it is dry where it should be dry and clean where it has to be clean. The principle also suggests that this aim should not be achieved by hiding everything under ground. Instead, as much water and water movement as possible must be visible, to make the city more attractive and the city dwellers more aware and involved.

4.3.4 Accentuating the area identity with water

Designing with water is not only a matter of shaping a programme and managing the water flows carefully and visibly, but of tailoring a plan to the local situation: to the *genius loci*. The hydrology of the landscape can be a source of inspiration for drawing up spatial plans and for the layout of residential and other buildings. Clean sources, drainage patterns, flood plains and underground water-bearing layers have an ecological potential that can be expressed in the plan. Doing so can contribute visibly to the identity of city and country, and to the quality and safety of the home environment, as evident in the rich variation in cultural history of the Dutch landscape.

4.3.5 Creating conditions for biodiversity with water

In the time when our technical resources were limited, human conduct adapted to the opportunities presented by the natural surroundings. This led not only to a wide variation in farmed landscapes, but also to great biodiversity. The ecological background is in the presence of what are known as gradients, which are gradual transitions of ecological conditions. One of the most important gradients is from dry to wet, which also has countless variations. Biodiversity is being put under great pressure now that technology is making it so much easier to control nature and to do almost the same everywhere. Designing with water offers a range of opportunities for putting interaction with nature first and restoring gradients that create conditions for great biodiversity.

4.3.6 Creating conditions for interactive processes

Water management has been organised traditionally both technically and hierarchically: top-down and to a very limited extent bottom-up. Experience with water planning in urban areas has led to more attention for interaction and dialogue with all the actors involved (Tjallingii and Van de Ven 2005). Not all parties need to be involved in the dialogue at all times, but an interactive process is necessary for realising and legitimising projects.

4.3.7 Creating conditions for an innovative learning organisation

Managing with simple standards is no longer sufficient in planning processes that involve multiple actors. Learning from experience in organisation and management is an important condition for dealing successfully with water. Jointly building up knowledge in a collective memory is a condition for integrated and sustainable water management.

4.4 GUIDING MODELS

These seven guiding principles set out the main direction, which can be summarised as more cooperation with nature and more cooperation between the actors. The next step is the development of practical conceptual tools for designers, which are referred to as the guiding models. They will be illustrated with reference to the Delft and Eindhoven water plans.

4.4.1 Three guiding models for Delft

An initial phase of the planning process for the Delft Water Plan involved a base study in which an analysis was followed by a forward-looking section with guiding principles and guiding models, and a section containing firm action items and programmes for sub-sectors (Tjallingii and Van Eijk 1999). The seven general guiding principles led in this case to three specific guiding models, which point the way for plans in various parts of Delft. These guiding models have been and are being used in elaborating the Water Plan and drawing up firm execution plans for parts of Delft, such as the post-war Voorhof and Buitenhof estates with blocks of flats.

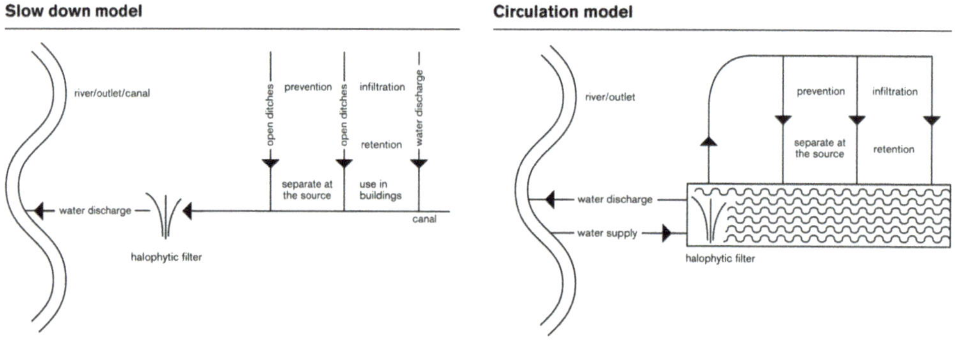

Figure 4.9.1 Guiding models for Delft.

Source: Sybrand Tjallingii.

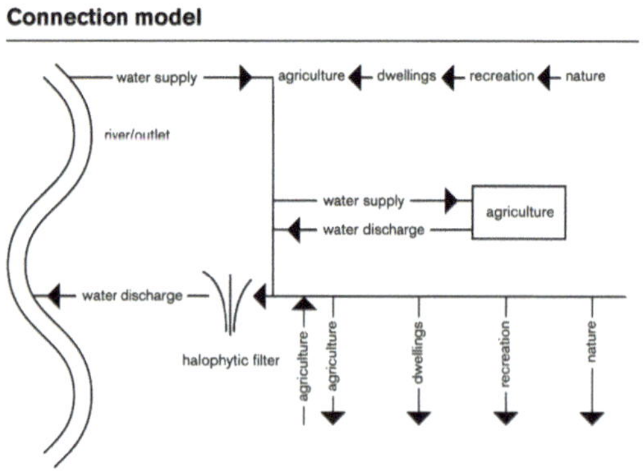

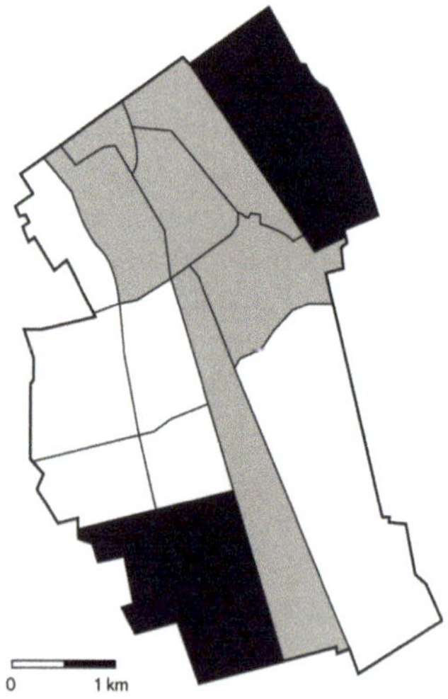

Figure 4.9.2 Map of Delft: the gray area represents the delay model, white the circulation model and black for the connection model.

Source: Sybrand Tjallingii.

The following guiding models were chosen.

- The *slow-down model*
  This model applies the principle of 'keeping clean and holding' through technical and spatial measures. The storm water is delayed and disconnected from the sanitary sewer. The model is in keeping with the Delft city centre and densely built districts with hardly any surface water and with mixed sewer systems. The measures are applied mainly on, in and around buildings.
- The *circulation model*
  This model introduces a circulation of surface water in addition to the measures on the building level. The water passes a cleansing reed or rush field (a helophyte filter). So much water is retained in the winter that it is unnecessary to let in *boezem* water in the summer. This seasonal storage may take place in lakes around the edge of the city. The system also covers peak storage. This guiding principle is compatible with districts with much public open space and with their own surface water system.
- The *connection model*
  This model is intended for the rural areas around the city. The district level circulation model is one of the links. Other links are nature conservation, leisure activity and agricultural areas. Water might run from one link directly into another, in *series*, but the links could alternatively be arranged in *parallel*, with the inlets and outlets kept separate, and the flows converging only downstream. This uses the flow principle 'from clean to polluted'. Another possible way of keeping clean watercourses in a clean state is by restricting foul discharges to one of them, so that the others stay relatively unaffected. Utilising the local hydrology is an important part of plans that follow this guiding principle.

4.4.2 Four guiding models for Eindhoven

The Blue Transformation is the name given to a radical change in water management and urban design (Tjallingii, Van den Top, Jonkhof, Wentink 2000). The municipality of Eindhoven drew up a handbook with this title based on experience gained in projects in the existing city. Guiding models give direction to this change process, with water as the 'vehicle'. A striking aspect is the construction of new water features, but the transformation of roofs, car parks and green strips also yields visible results. The urban landscape that ultimately emerges after disconnecting storm water from the sewer system blends the built area, green structure and water features into a single entity. The planning and design of this entity connect with the issues on several levels.

The following two guiding models were used in Eindhoven alongside the slow-down model and the connection model discussed above.

- The *infiltration model*
  The infiltration model focuses on the infiltration of water into the soil as well as measures on building and street levels. The soil frequently consists of sand, and the ground water table is relatively low. A situation of this kind is the exception in Delft, but is the rule in Eindhoven. Infiltration tops up the groundwater, which is important in view of the falling groundwater tables in these areas caused by the

increasing paved area, drinking water extraction and agricultural level control. What the model advocates in the urban district is rain water infiltration in swales that are usually dry, but are able to take the water from several days of heavy rain, which then slowly soaks into the soil. In Dutch practice, these swales are known as *wadis*, and they can be spatially integrated into green strips. The guiding model

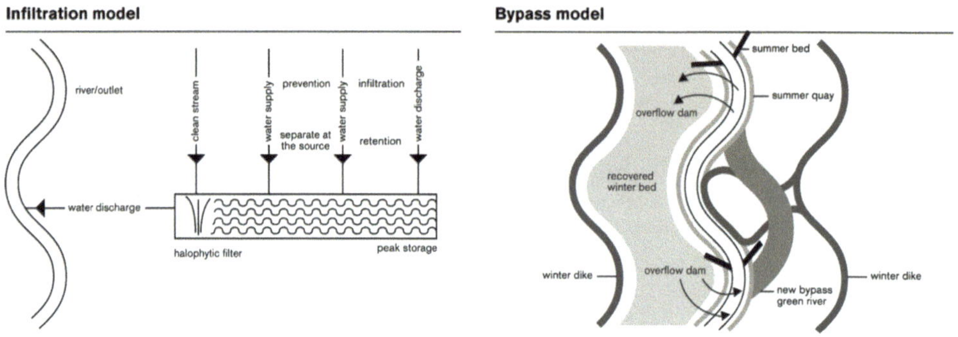

Figure 4.10.1 Two extra guiding models for Eindhoven.

Source: Sybrand Tjallingii.

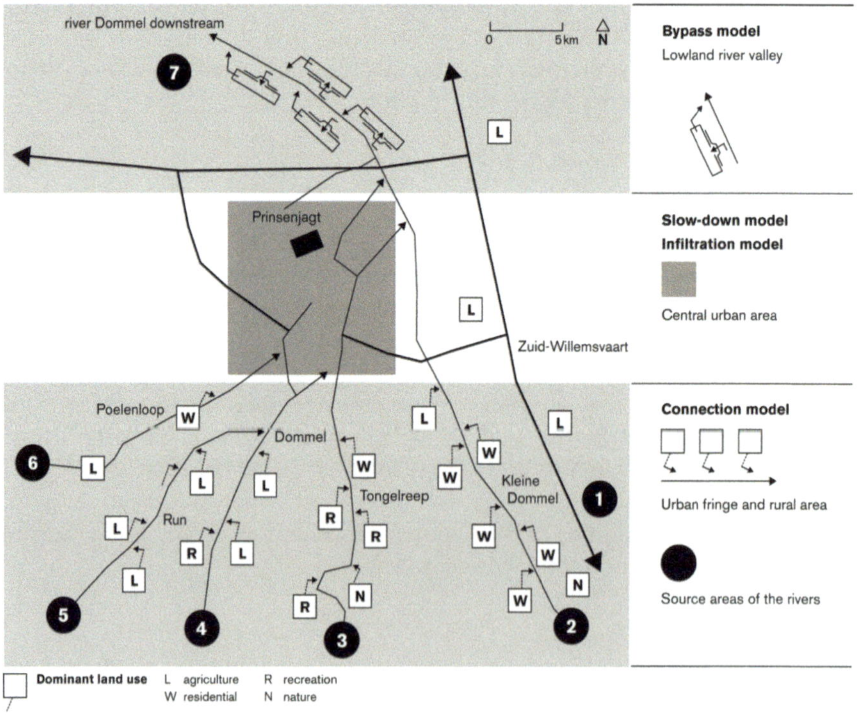

Figure 4.10.2 Working with guiding models in Eindhoven.

Source: Sybrand Tjallingii.

also suggests a peak storage pond. The water drained from the urban district is purified as it passes through a helophyte filter on its way to the river or stream.

- The *bypass model*
  The *bypass* model is appropriate for water management in a stream or river valley, where the increasing paved area and the periodic precipitation peaks sometimes cause considerable nuisance. Preventive efforts such as retention (holding within the surface water) and infiltration (holding within the groundwater) notwithstanding, the river valleys downstream of urban areas will also have peaks of this nature to deal with in the future. However, the riverbeds are often narrowed by quays and buildings. The guiding model suggests enlarging the flood plain, which can be found both inside and outside the old riverbed. Sometimes a bypass in the form of a 'green river' may offer a solution.

4.5 WATER ISSUES IN THE PLANNING PROCESS

Urban water management is more than sewers, gutters and pumping stations alone. Water management demands cooperation with nature and cooperation between the various actors. Urban water plans are duly being drawn up to link the layout of urban areas with the water issues. Water adds to the beauty of cities, but too little, too much, or too dirty water also creates problems. The climate is changing, and towns and villages are continuing to expand. Time is pressing, and the 2003 National Administrative Agreement on Water obliges all municipalities to draw up an integrated and sustainable urban water plan. Much has already been written on the necessity and content of the water plans, but many questions also surround the planning process. How can the water issues be incorporated into the planning process?

How does an urban water plan take shape? The authors of *Water in Drievoud* (*Water in Triplicate*) analyse the experience with three working methods: the target image, the guiding principle and the negotiation approaches (Tjallingii and Van de Ven 2005). The definition of the water issue differs in these three approaches. The first phase of the target image approach defines a final state, which is then aimed for. The guiding principle approach formulates the issue as guiding principles and guiding models for orienting the design, with the execution plan and execution objectives appearing only in the final phase. The negotiation approach opts for an open planning process in which the direction and the final objectives depend primarily on the consensus established between the participants. The three approaches are different but the analysis of *Water in Drievoud* also demonstrates how the three can supplement each other. The guiding principle approach is important mainly in the strategic phase of planning processes, leaving target images to the operational phase. The process is not completely open, but is driven by the need to give water issues an appropriate place. Moreover, lessons learned in earlier plans guide the present planning process. Guiding principles and guiding models embody these lessons. Together, they form the toolkit for planning with water.

Chapter 5

More water in the historic city centre: transformation

Wout van der TOORN VRIJTHOFF and Erwin HEURKENS

Department of Real Estate and Project Management, Faculty of Architecture, Delft University of Technology, Delft, The Netherlands

5.1 INTRODUCTION

The urban fabric of city centres is fine grained, with narrow streets, dense building and usually little surface water. Access to the centre, where there are large-scale functions or functions that serve a large area, is therefore an important urban design issue. This aspect may conflict with a desire to restore filled-in canals, with a view to making the city more liveable and creating more surface water, often at the expense of access roads and parking spaces. Nonetheless, more surface water will be appearing in many city centres, such as Utrecht and Breda, in the coming years.

Other water issues for the city centre are retaining storm water and disconnecting the sewer network (part of the first guiding principle). Disconnection is not always possible as such, but when combined with a transformation issue, both often become feasible. For example, new buildings can be designed to retain storm water for longer.

City centres are often situated on higher ground, but being located on rivers or in the polder landscape means they can still be exposed to flooding. Additional investment in technical and spatial means of retaining storm water (which is known as the delay model) is called for where this problem exists, as in Delft city centre.

5.2 CATHARIJNESINGEL, UTRECHT

5.2.1 The area

Utrecht came into existence approximately 2000 years ago as a Roman army camp. A start was made on digging a new defensive canal (*singel*) around the city after it was granted a charter in 1122. The *singel* was preserved intact even after the 1874 Fortifications Act. When the *singel* lost its significance as a navigation channel around 1970, the northwest part (approximately one-quarter) was filled in to make room for roads and parking spaces. However, the ring road system was never completed, and a substantial area of new asphalt has only ever seen limited use.

Car use has now increased so much that different means of transport are needed to prevent gridlock in the city centre. There is also a growing need to make the surroundings of the filled-in *singel* a pleasant place to be and to restore the human scale, or, in other words, to improve the livability. The relatively clean surface water in the city is once more a popular part of the public space. All the above has prompted the city council, with the support of the population of Utrecht, to reopen the filled-in waterways and restore the structure of the old *singel*.

Figure 5.1 Utrecht.

Source: Topografische Dienst Kadaster, Emmen.

5.2.2 The water issue within the urban water plan

The primary objective of the Utrecht Water Plan, which was drafted following the report of the 21st Century Water Management Committee (WB21), the adoption of the European Union Water Framework Directive and the national debate on optimising the water chain, is a water system for the future, in keeping with the natural conditions on and

Figure 5.2 Restoration of the Utrecht *singel* structure.

Source: Municipality of Utrecht.

around the Utrecht territory. The spatial structure offers opportunities to achieve this aim, and even to strengthen it. The Water Plan has the following ambitions and starting points:

- to contribute to retaining or creating an attractive living and working environment;
- to strengthen the Utrecht water system, in keeping with the natural conditions on and around the Utrecht territory;
- to manage the water system and the water chain cohesively and efficiently, and in a way that is clear for the public, oriented to multiple forms of human use and natural assets;
- to fit in well with the urban dynamic;
- to coordinate water and spatial planning effectively;
- for water to be a source of quality for the city.

The Water Plan (2005), together with the Regional Structure Plan (2004), the Municipal Structural Concept (2004) and the Environmental Policy Plan (2004), form the water framework within which operational plans will be drawn up.

The water managers and owners, who are known as the water partners, have jointly drawn up the Utrecht Water Plan. The cooperating parties are the Higher Water

Figure 5.3 Chart showing the Utrecht Water Plan, medium-term view.

Source: Municipality of Utrecht.

Boards De Stichtse Rijnlanden (HDSR) and Amstel, Gooi and Vecht (AGV), the water distributor Vitens, the provincial government of Utrecht, the Directorate-General for Public Works and Water Management for Utrecht and the municipality of Utrecht. The Directorate-General for Public Works and Water Management controls the Amsterdam-Rhine Canal and the Merwede Canal, the HDSR controls the primary and secondary watercourses and the wastewater treatment plants, and AGV manages the water in the river Vecht and the Klopvaart waterway. The smallest watercourses, the capture of storm water on street surfaces, the sewers and the qualitative management of most groundwater are the responsibility of the municipality of Utrecht. Hydron Midden Nederland extracts, purifies and distributes drinking water and process water. The provincial government of Utrecht is responsible for the quantitative management of all the groundwater beneath Utrecht and for the qualitative management of the groundwater in the groundwater protection areas.

Utrecht is a water management and wet infrastructure hub in the west of the Netherlands. Its location at the end of the river Kromme Rijn (or 'Crooked Rhine') area and its extensive paved surface make the city vulnerable to flooding, in theory. Flooding does occur incidentally in practice, but is not a major problem, perhaps unlike the downstream areas. Utrecht's freedom from the risk of flooding depends on several factors, including the elevation of the city, how sensitive the infrastructure and public life are to excess water, the surface water and groundwater levels, and future climate change. A city is normally extremely vulnerable to a rise in precipitation levels or incoming surface water volumes.

The water issue in the Utrecht Station Zone project falls within the public space category. One of the aims of the Station Zone Structure Plan (February 2006) is to improve the connection with the city centre. The principal objective is to create a new centre for Utrecht by integrating the upgraded station zone with the city centre, to their mutual benefit. The link between the two zones must be made in the public space.

The restoration of the *singel* structure is an essential part of the plan. It is one of the drivers of large-scale change within the station zone. The Catharijnesingel will act as the go-between for two now separate worlds, the old city centre and the Hoog Catharijne retail complex. New links are being created between Vredenburg square and the Catharijnesingel. Vredenburg; on the east side of the *singel*, will be a social, recreational and shopping area, with apartments above the shops. The Music Palace, the new Hoog Catharijne and the V&D department store will also be accessible from the *singel* side, thus ensuring effective interaction between the buildings and life on the street and the *singel*. The Vredenburg side will be a waterside route for cyclists and pedestrians. The main route for road vehicles and cyclists is to the west of the water, towards the station, where street-accessed non-retail commercial premises, service companies and possibly shops will be based.

The Utrecht municipal administration has decided to restore the part of the *singel* that is now filled in to strike a balance between roads and car parks on the one hand and the city centre with its restricted traffic access on the other. The main point is the restoration of the centuries-old water structure, and has little to do with nostalgic design.

Consultation has taken place with the Water Board Stichtse Rijnlanden within the framework of the Water Assessment that started on 1 November 2003, to draft a joint water clause for the Station Zone Structure Plan. The clause describes the existing situation (as at February 2006). The station zone has a considerable paved surface in the form of infrastructure, car parks and large contiguous roof areas. The existing water system in the station zone is based on rapid drainage. Storm water is discharged directly into the groundwater, sewer system and surface water. The wastewater is transported under normal circumstances to a treatment plant. No additional storage capacity, with the exception of the combined sewer system, is present in the station zone, which means that large precipitation peaks cannot be stored in the zone itself. The surface water quality in the station zone is mediocre because of sewer system overflows and various sources of pollution, such as road traffic, the bus station and the railway yard, and discharges from treatment plants. The biological and ecological water and bank quality in the station zone is minimal.

It has been decided that the station zone development will aim to intensify the use of space. A common incentive for a choice of this kind in restructuring city centres is the need to generate revenue. However, it is particularly difficult to increase the storage capacity in areas of this kind. Nonetheless, the structure plan is an improvement, because the amount of open water in the station zone will increase from 1% to 3%, and some 95% of the urban function expansion will be realised within the existing paved area.

It would seem obvious that the structure plan is unable to offer sufficient space to remove all possible obstacles within the station zone. The proportion of open water needed

Figure 5.4 The Utrecht water system: operating on the regional, city and district scales.

Source: Municipality of Utrecht.

to achieve the drainage factor desired by the Water Board Stichtse Rijnlanden (which goes unmentioned) is estimated at 13%, which is not realistically feasible in a central urban area.

However, the economic damage caused by excessive rain in the station zone may be considerable, and therefore the parties directly involved within the framework of the Water Assessment have defined joint ambitions and explored possible measures. An assessment will be made of the necessity of actually implementing these measures in the production of the sub-plans and definition of the sub-projects. Factors to be taken into account will include the social feasibility of the measures and the trade-off between the costs of the measures and the costs of possible emergencies.

5.2.3 Solutions and process planning

The twin objectives of the Singel Structure Restoration Project are to restore the *singels* (in a modern form) as a landmark element of the city, and to boost the quality of

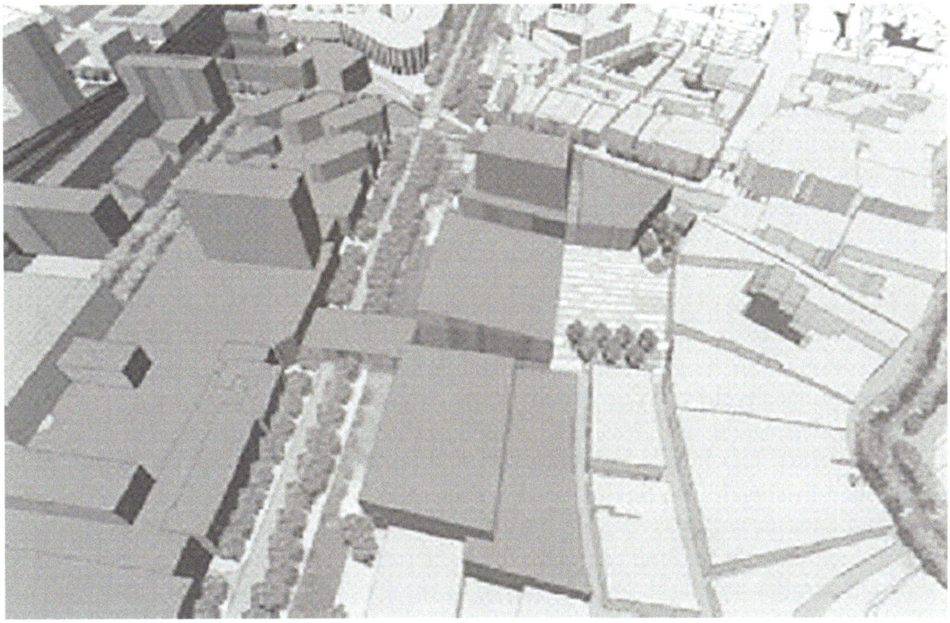

Figure 5.5.1 Artist's impression showing the strategic position of the Catharijnesingel for both the station zone and the city centre.

Source: Municipality of Utrecht.

Figure 5.5.2 Artist's impression of the public space on the Catharijnesingel.

Source: Municipality of Utrecht.

the surrounding areas and the city as a whole. The aspects involved alongside livability include historical, environmental, tourism-related and commercial. In addition to these aspects, countless elements have been incorporated into the restoration for adding value to the zone and increasing its appeal, which are known as the 'pearls'. Among the examples are the remains of an ancient outlying defensive wall and a section of old city wall with defensive towers, the construction of a sustainable ecological corridor, provisions for bats, moorings for pleasure boats, sanitary facilities, pavement cafes, assorted greenery and specially designed jetties to encourage contact with the water. An innovative element is the approach to bank vegetation and locations for new wall vegetation.

The restoration of the *singel* structure will extend the ecological infrastructure for natural assets in the city. The main function of the *singels* is as a corridor running from the Kromme Rijn via the Catharijnesingel to the river Vecht. A problem that is generally recognised is the pollution from sewer overflows in heavy rainfall.

An underground link was built when part of the Utrechtse Singel was filled in around 1970. This link, which consists of a conduit of 3 by 5 metres, runs parallel to the filled-in *singel*, and has ensured sufficient flow through the *singels*. The underground conduit will lose its function once the *singe*l has been restored. Restructuring plans include underground car parks for large parts of the Catharijne zone. The underground conduit will have to be removed in the parts of the plan area concerned. The underground conduit can be reused for water storage in places where removal is unnecessary. An example is the section running adjacent to the Moreelse Park and the Mariaplaats Square in the south of the plan area. This measure can create 3525 cubic metres of water storage.

The municipality of Utrecht's priority for the water issue is the quality, in particular by avoiding overflows. This aim is hampered by a combined sewer system in the city centre. The majority of the excess water during torrential rain is discharged directly through the system. If the discharge capacity is insufficient, the overfalls come into action and a mixture of storm water and sewer water ends up in the surface water, which is thus polluted. However, there is no solution to this problem for the city centre in the short term. As and when construction or renovation is carried out in the city centre, the municipality insists on separate connections for every building for clean rainwater and sewer water to a drain below street level. It will then be possible later to introduce a separate sewer system in the city centre. The clean water from city centre buildings situated directly on a canal or *singel* can be discharged directly into the open water, which forms a separate system on a very modest scale in the city centre. The underground water storage will be used first as overflow, avoiding some of the spillage into open water.

Reintroducing the *singel* structure actually means enlarging the surface water by widening the profile, which is also compatible with the objective of enhancing the liveability of the city centre. The link between the Catharijnesingel and the Leidsche Rijn waterway also reinforces the quality of the surface water by improving the through flow of water. A pump is available at the Weerdsluis lock for pleasure boats to discharge their wastewater into the sewer, and bank vegetation is planted at various places on and under the water berm. This may make a small contribution to the surface water quality.

5.2.4 What does it cost and who pays?

The idea of restoring the *singel* structure within the framework of the Utrecht Centre Project (UCP) development was as a lever to generate support for the plans. The municipality used the concept as a showpiece for the UCP. As a result, the restoration of the *singel* would appear to have largely symbolic value. A concept that figured prominently through the years of political debate surrounding the UCP was liveability. More offices and car parks, in the context of the debate, made no positive contribution to liveability in the zone, unlike a restored *singel* structure, which therefore ensured balance in the UCP plans. This balance appeared essential for approval from the local politicians. Against this background, the contribution made by the restoration of the *singel* structure to solving the urban water issue was a relatively minor matter.

The *singel* project has a clear funding structure. The municipality's primary objective of improving the living climate in the city centre is still an obstacle to direct investment from the Water Boards De Stichtse Rijnlanden (HDSR) and Amstel, Gooi and Vecht (AGV). However, the Water Boards nonetheless set conditions on the restoration of the *singel* structure as part of the water system. The issues of concern include the influences and impacts of solutions on the through flow of water. It has been agreed that the Water Board will manage the *singel* after completion, which implies that the Water Board will be responsible for regular cleaning and maintaining the depth of the water. The municipality will manage the banks.

The municipal budget for the project expenses is currently less clear. The restoration of the *singel* structure was estimated in the 1990s at 100 million guilders. After conversion and allowing for inflation, the amounts for the entire project would be approximately 50 million euros. This includes a cost item for the public space in sub-areas situated alongside the *singel*. Two important building plans can be distinguished:

- the construction of the public transport terminal and surroundings;
- the restoration of water in the Catharijnesingel and the redevelopment of the surrounding areas (Vredenburg and Gildenkwartier).

The peak of the activities is in the period between 2006 and 2010, when the municipality of Utrecht will be working simultaneously on the public transport terminal, the Catharijnesingel, the surroundings and the main infrastructure. The next operation will be started only when the activities on the previous site are complete. The municipality and partners will arrange for alternatives during the construction to enable shops and businesses to receive supplies and remain accessible. The municipality will strive to minimise the nuisance and will keep a close watch on traffic changes resulting from the activities. It is difficult to give a sound estimate of the costs of restoring the *singel* structure because it is so tightly interrelated with other components of the total UCP. Compared with the costs incurred for restoring the harbour in Breda, 50 million euros would appear to be a reasonable guideline. This amount contrasts sharply with the costs to be incurred in creating new watercourses outside the city centre area.

The municipality is provisionally setting 2015 as the end of the final phase of the *singel* project. However, it should be noted that the plans depend on the land developments of

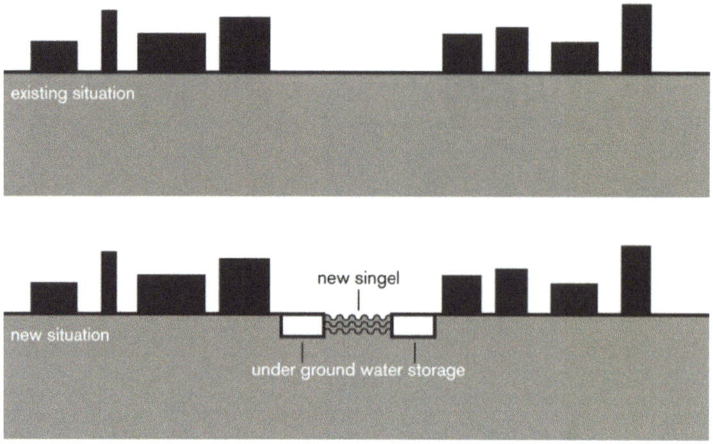

Figure 5.6 Existing and new situation in Utrecht.
Source: Wout van der Toorn Vrijthoff.

the station zone sub sectors. The complexity of this project means that parties have had to assume multiple starting points for the ambitions for the station zone, which has delayed much of the *singel* project. Future developments for the station zone therefore largely determine the duration of the *singel* project, although the support for a 'liveable singel' is currently established among the municipality and residents. Furthermore, contractual agreements have been entered into on this subject between the municipality and station zone developers.

5.3 OLD HARBOUR, BREDA

5.3.1 The area

Throughout most of its history, the harbour in Breda has been functional, and one of the reasons for the city to exist. More than 700 years after Breda received a city charter (in 1252), a decision was made in 1964 to fill in this 'monument' to improve access to the city by car. The course of the river Mark originally formed the west city canal and the boundary of Breda. Breda harbour was the wide section of the river Mark. Plans were drawn up as far back as the early 1930s for a broad highway between the south and the north of the city. This 'corridor' was intended to improve access to the city and followed the course of the river Mark along the west side of the old city. What this meant for the harbour zone was the construction of a large multi-storey car park under the four-lane road, which was built after the harbour was filled in.

Plans for restoring Breda's filled-in inland harbour started sometime in the 1990s, prompted by a growing interest in cultural heritage and the policy of restricting traffic access to the city centre. A broad debate gathered momentum on a local level, culminating in firm plans for restoring the harbour.

Figure 5.7 Breda.

Source: Topografische Dienst Kadaster, Emmen.

Breda municipal council approved the 'provisional outline design' for restoring the old harbour in September 2000. An important idea underlying the decision to bring the water back was to boost the city's experiential value. The objective is a visual impact that appeals to residents' and visitors' sense of history. The municipality made a budget of 29.5 million euros available for bringing back the water to the harbour. This substantial investment in the public space creates favourable conditions for private investments in adjoining existing property and new building development on undeveloped areas in the immediate surroundings. The municipality's aim is to fulfil the plans completely by the end of 2008.

Breda is not unique with its plans for restoring historic watercourses in the old city centre. Den Bosch was a pioneer in this area. The restoration of the Binnendieze heralded a reversal from a policy oriented to renewal to one dominated by preservation and restoration of the historic city centre. Many cities in the Netherlands and elsewhere are developing plans with the same intentions. The European Union is duly supporting a joint initiative of six European cities for exchanging experiences and developing knowledge. Breda is the initiator and coordinator of this project.

5.3.2 The water issue within the urban water plan

The draft Breda Water Plan is a comprehensive plan of the municipality of Breda. It was drawn up with four water partners: the water distributor N.V. Waterleiding Maatschappij Noord-West-Brabant, the District Water Board Mark en Weerijs, the Water Board for

Figure 5.8 Activities for opening the harbour.

Source: Municipality of Breda.

West Brabant and the provincial government of North Brabant. The new water management objectives in the 21st century are translated to the Breda situation in the water plan, in which the partners have devised joint target scenarios for 2030. The scenarios were compared with the existing state of the water system and the water chain, following which alternative solutions and measures were drawn up, and binding agreements for the future entered into between the water partners.

In the light of the recent insights, the natural system of Breda is being taken as a starting point for restoring the water system. It is hoped that this will achieve greater resilience to changes in the system, and likewise to enhance safety, health and the experiential value.

The target scenarios for 2030 are grouped under four topics:

1. 'estates along streams and river' – aspects of experiential value, natural assets, leisure activities, resilience and water quality;
2. 'water in the neighbourhood' – aspects of health, responsibility and aesthetic value;
3. 'water hub Breda' – aspects of safety, layout, maintenance, resilience and water quantity;
4. 'water partners' – subjects such as water as an ordering principle, cooperation and coordination.

The opening of the harbour is included as a water issue in the various topics as follows.

- Theme 1. 'Estates along streams and river'
  Water and cultural history have united and reinforce each other in the central area of Breda. To the north of Breda too, the historic moats and forts will be made more visible by restoring water access to the area, thereby reinstating something of the old function as part of the historic Waterlinie defence line.
- Theme 2. 'Water in the neighbourhood'
  Water of good quality is properly visible in the neighbourhood in a friendly variant, the most important function of which is 'perception'.
- Theme 3. 'Water hub Breda'
  The opportunity to lead some of the water flow through the centre, which was presented by opening the harbour, was taken, thus enhancing the discharge capacity.

Thirteen project programmes are set out in the Water Plan starting from the formulated target scenarios and the projects and measures already planned. One of the project programmes is the opening of the harbour, which will bring water back to Breda city centre within the framework of the Oost-Westflank project.

5.3.3 Solutions and process planning

The filled-in harbour was the subject of intense debate in Breda. The decision to satisfy the demands for accessibility in support of the economic interests in the historic city centre was in keeping with the spirit of the time. A powerful countermovement that came into being at the same time argued for preservation and restoration of the cultural heritage that is inextricably linked with Breda's origins. The discussion surrounding the decision to fill in the harbour waxed and waned for decades, influenced by changing social circumstances, such as:

- growing interest in cultural heritage;
- public recognition of the importance of cultural heritage for the attractiveness of the historic city centre;
- support for this public recognition from economic interests, related to the significance of the historic city centre for tourism;
- a moderation of the demand for road access to the historic city centre;
- the fact that certain investments were functionally and economically depreciated.

Ashok Bhalotra's original suggestion in 1994 for bringing back the water in the Westflank district of Breda city centre met with broad support. Some of the support came from the view that filling in the harbour was a historical error. The resolution to fill in the harbour was passed in the council meeting in 1964 with a majority of one vote. The considerable resistance to filling in that started in the early 1960s has persisted ever since. Another element of the broad support was based on the low experiential and use value of the public zone near the filled-in harbour that existed at the time. A round of talks conducted by the consultancy BRO with local organisations and directly involved parties revealed the following negative aspects of the existing situation in Westflank:

- an extremely poor environment for pedestrians and cyclists;
- conspicuous and pervasive road traffic;
- an unintelligible structure;
- a mediocre public space layout;
- the stark contrast with the recently redesigned city centre.

Soon after the idea for reconstruction was launched, the question changed from whether the water would return to what form it would take. A whole series of constraints came into play in the search for an answer. The most important objective, but not the only one, was the restoration and preservation of cultural heritage, in support of the attractiveness of the historic city centre. The 'Starting points memorandum on the provisional outline design for Oost-Westflank' of June 2000 contained the following objectives:

- to strengthen the economic structure of the west side of the city centre;
- to promote tourism and leisure activities;
- to strengthen the residential function in the city centre;
- to strengthen the spatial and functional relationship between the station surroundings and the city centre;
- to strengthen the qualitative constraints for the approach to urban renewal in the west part of the zone inside the *singels*;
- to reassess the cultural and historical heritage;
- to improve soil quality;
- to improve the spatial quality and the everyday living and working environment;
- to ensure integrity of development and management of the water on various scales as an extension to the ecological network and the water chain;
- to develop and manage the public space sustainably.

Most of the above objectives can be placed under the principal objective: to increase the attractiveness of the historic city centre. However, this is not bound to lead to opting for restoration of the harbour and the old watercourse. The desired outcome alluded to in the objectives could possibly have been achieved through different spatial interventions.

Agreement has now been reached in Breda on:

- the importance of restoring the oldest monument in the city;
- the significance of doing so for the city's identity and competitive position, as manifested in such things as sales growth trend forecasts, especially in the city centre retail and catering sectors (based on: BRO 2000);
- the significance for the increase in land and property values in the immediate vicinity of the harbour.

On this basis, private developers were willing to waive part of the added land and property value to cover the costs of the intervention in the public space (equalisation). At the same time, discussion is still ongoing on the authenticity aspect. It is debatable whether the monument should be restored to its original state. A combination of technical and financial considerations has led to a decision that 'look alike authentic' is good enough for parts of the plan.

The urban water issue had yet to become an item for discussion when the plans for restoring the harbour were made. The starting points for the provisional design therefore contained no objective for improving urban water management. The district water board was consequently scarcely involved in the planning. The first talks with the district water board were in 2003, and one of the objectives was to agree a financial contribution. The district water board assessed the main impact of the plans for urban water management to be a better through flow in the surface water network in the city centre, which is an improvement in the surface water quality. However, the district water board was unwilling to make a substantial contribution at that stage in the project development, bearing in mind that the decisions on funding the plans had already been made.

There is usually much discussion of authenticity in planning the restoration of historic urban watercourses, but concessions are often made nonetheless, in particular under pressure of financial feasibility. The need for identity is apparently sufficiently satisfied by something that evokes associations with the current public image of history.

This discussion is an issue in Breda too. It has been decided to restore the harbour that was the original reason for Breda's existence. However, this is occurring in the twenty-first century city. For instance, allowances are being made for the traffic caused by visitors and supplying businesses, and consequently the water will be narrower by 3 to 4 metres. Furthermore, the quay will have both high and low levels, to optimise the recreational function, but with due regard for the old town and quay walls.

Breda therefore does not want to put the water back exactly where it once was. Furthermore, the old harbour also went through various phases: water with banks that overflowed when the water level was high, water with a quay on one side, and a harbour with cargo vessels. What constitutes authentic is an open question. The final choice was not for a quaint harbour in Anton Pieck style, but one ready for modern

Figures 5.9.1, 5.9.2, 5.9.3 Breda harbour, then, now and in the future.

Source: Municipality of Breda.

use, where events can be held, where vessels can moor and where boat tour passengers can embark and disembark. Breda believes that with a modern variant of the harbour it can restore the identity of a city centre on historic water.

5.3.4 What does it cost and who pays?

The public funds available to municipalities fall under three categories:

1. own resources (municipal taxes and duties), amounting to approximately 17% of income, one-third of which from property tax (OZB);
2. the general contribution from the Municipalities Fund, amounting to approximately 37% of income;
3. specific contributions, which are approximately 46% of income.

The first two are part of the autonomy of the municipalities, and the third comes under the heading of 'joint administration'. The first category includes property tax

Westflank Breda 'The new river'	Costs	Coverage
Coverage costs reconstruction inner city historical water	**29.500.000**	
Budget municipality		
General subsidy municipality fund		20.415.000
Local taxes		0
Compensation fund VAT taxes		4.710.084
Private sector		
Ground exploitation new buildings (indication)		2.000.000
Exploitation agreement real estate		0
Contribution water board		25.000
Contribution state funding		1.650.000
Contribution EU funding		700.000

Figure 5.10 Resources for covering Westflank costs.

Source: Municipality Breda, Wout van der Toorn Vrijthoff.

(OZB), which accounts for about one-third, plus the betterment levy, second-home tax, tourist tax, parking taxes, dog licences, advertising tax and the tax on encroachments on or above public land.

The municipal contribution to the Westflank costs must be covered by the first two categories. The assumption is that the municipal land agency budget forms part of the municipal budget.

The plan for bringing the water back to the Westflank district of the old town, popularly known as the 'New Mark' construction, consists of several components. Restoring the old watercourse, including the harbour, is a restructuring intervention on part of the urban public space, which will be accompanied by redeveloping or refurbishing adjacent existing buildings and implementing building plans for undeveloped areas.

Plans for reorganising the public space predated the decision to reintroduce water into the Westflank district of the city centre. The investment necessary for implementing these Westflank reorganisation plans was estimated at € 16 million. Work on plans that had restoration of the harbour and the Mark clearly in scope proceeded in the 1995–1997 period. The base model for these plans was introduced into the public participation process in 1998. This base model was subdued and functional, and the scale of water in the plan area was duly limited. A municipal council resolution adopted in September 2000 made a budget of € 23 million available, based on a provisional design. However, the plans continued to develop, partly in response to the ongoing interaction with interested parties, organisations and engaged members of the public. The plan for water in the Westflank district then became more ambitious, with the 'Case for Character' plan, which had been presented back in October 1999, as reference. This plan for the 'New River' was presented by HeJa project development and NBM Amstelland, with a dominant and inspiring contribution from Eloi Koreman. The discussions on the ambitions, in which model 4+ and model 5 featured significantly, led to a transformation of the provisional design into multiple variants. The most important differences between model 4+ and model 5 are:

- more respect for the original character of the water in the form of a river, rather than a tightly controlled system of canals;
- where there is a river, there also have to be buildings that rise out of the water and form parts of the quay wall, of which the building 'Op de Trapkes' is a historic example. Model 5 adheres more closely to the original building lines;
- the broadening of the harbour to almost the original dimensions.

The costs for models 4+ and 5 far exceeded the budget set in September 2000 based on the provisional design. The 2002–2006 programme agreement then set a normative investment budget of € 29.5 million, without reference to the technical specification of the provisional design or the variants developed later. Figure 5.11 includes the figures stated above.

The budgetary framework necessitated a combination of cutbacks and a tempering of ambitions. The main element of the retrenchment was a new plan variant for raising the river between the Trapkes and Karnemelkstraat by 1.70 metres. This variant has the following impacts:

- less excavation depth;
- shallower dam walls that require no anchoring;
- the absence of brick cladding for the quay walls, only a very small part of which are visible;
- the main sewer can continue to run under the river;
- a culvert under the Markendaalseweg instead of a bridge.

	Million	Price level	Index
Dry Westflank sollution	€ 16	1997	100
Counsil decision Sept. 2000 (about water in the Westflank)	€ 23	1997	100
Calculated model 4+	± € 32	2001	120
Model 5- city river	€ 35.9	2001	120
Approval of the program 2002-2006 (agreed budget)	€ 29.5	2002	125

Figure 5.11 Costs of redeveloping the Westflank public space, consisting of the harbour and Markendaalseweg Noord.

Source: Municipality Breda, Wout van der Toorn Vrijthoff.

	Cost estimation		
	2000 Starting-point	**2002 Agreement program**	**2002+**
Simple version without water in the Westflank	x (20.6 million)		
Community participation version with limited water	x+7 million (26.2 million excl. plan development)		
Model 4+	x+ 8 million (28.8 million)	INDEX 3% → 29.5 million	? → 35.8 million
Model 5	Not calculated		

Figure 5.12 Estimate of the costs.

Source: Municipality Breda, Wout van der Toorn Vrijthoff.

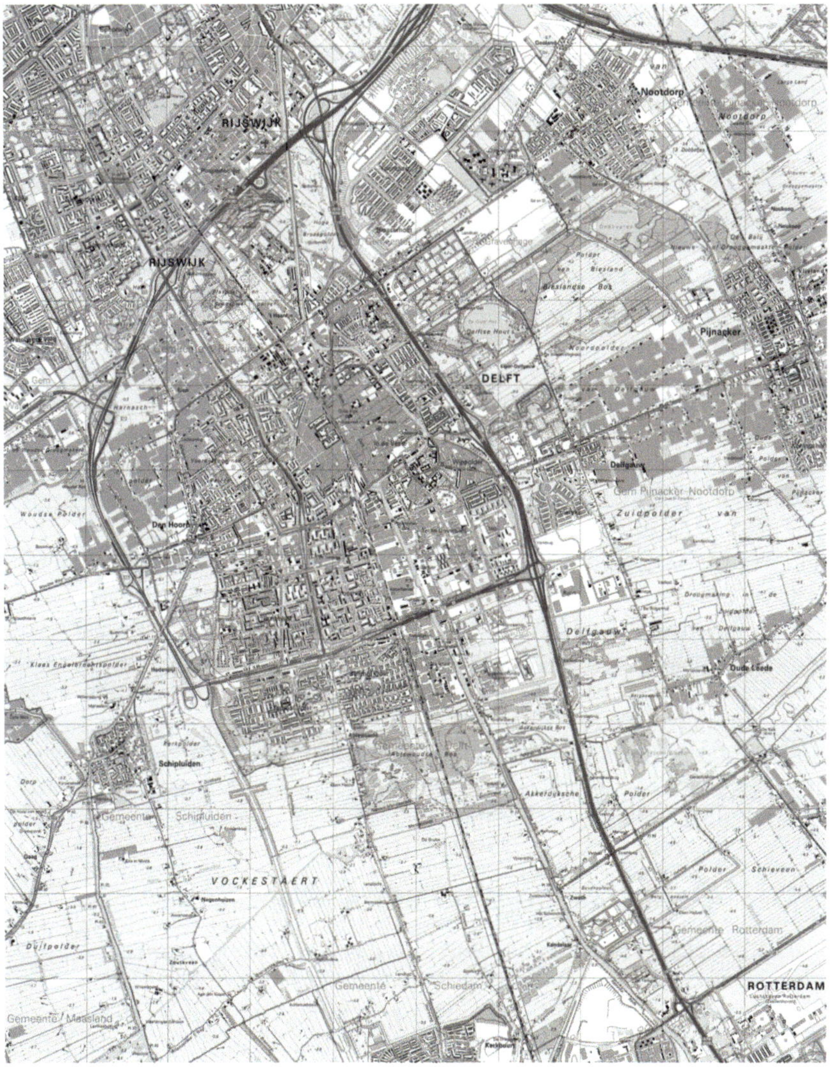

Figure 5.13 Delft.

Source: Topografische Dienst Kadaster, Emmen.

5.4 EAST CITY CENTRE, DELFT

5.4.1 The area

Delft was built on a dug watercourse, the Delf, and its name is indeed derived from *delven*, which means to dig. A count's manor was situated on the elevated site where the Delft crossed the creek wall of the silted up river Gantel, probably since the eleventh century. This is one of the reasons why Delft was an important market centre, evidence of which is still visible in the size of the central market square. The river Gantel runs through Monster and Poeldijk in the Westland region. The Gantel was once a broad creek that was part of a complex system of tidal creeks running into the Maas estuary.

Several centuries B.C., the North Sea broke through the bank of the river Maas near Monster. Almost the entire Westland region was then inundated. Many tidal creeks came into existence in the peat area, including the Gantel, with sand and clay sedimentation. Over several centuries, these strips of firmer soil rose above the surrounding peat, as it dried and subsided. The tidal creeks silted up completely during and after the Roman period. The area running from Monster past Poeldijk and Wateringen to Delft and Pijnacker was therefore covered with fertile clay and sand, which was known as Ganteldek. The sand and clay ridges so created were suitable for building houses, and were sought after for inhabitation.

Delft has a historic city centre with a fairly straightforward street plan. The Old Delft and New Delft canals run parallel to each other in a roughly north-south direction. The New Delft canal is better known as the sequence Koornmarkt-Wijnhaven-Hippolytusbuurt-Voorstraat.

With the levelling of the city walls in the nineteenth century and the arrival of the train in 1847, Delft again became an attractive site for new industries, such as the yeast and spirit factory (later Gist Brocades, which is now part of DSM), Calvé and Delft Instruments. The foundation of the Royal Academy (later to become the Technical University) in 1842 and the Netherlands Organisation for Applied Scientific Research (TNO) in 1932, meant that Delft also became a centre of science.

Delft expanded considerably in the 1960s, especially to the south, where the high-rise estates Poptahof and Voorhof, which are still among the most densely populated estates in West Europe, and the somewhat less ambitious Buitenhof, were built, in that order. The Tanthof estate was developed even further south starting in the 1980s. Tanthof East is a well-known exponent of what is called the 'new drabness', with an untidy street plan characteristic of estates of its kind. Tanthof West is a somewhat better organised estate of single-family homes situated on the Derde-Werelddreef.

These developments gave Delft an extra centre alongside the historic city centre, in the form of the In de Hoven shopping centre on the opposite side of the rail line.

5.4.2 The water issue within the urban water plan

The Delft Water Plan, entitled 'A Blue Network', was adopted on 27 April 2000. This plan outlines water management objectives for the coming years. The plan also includes some concrete measures, such as those for preventing flooding in Delft city centre. Delft suffered severe flooding in 1998 and 1999. The municipality of Delft together with the Water Board for Delfland commissioned the study 'Future survey of the water and soil levels of Delft city centre'. The conclusion was that the quays in the east of Delft city

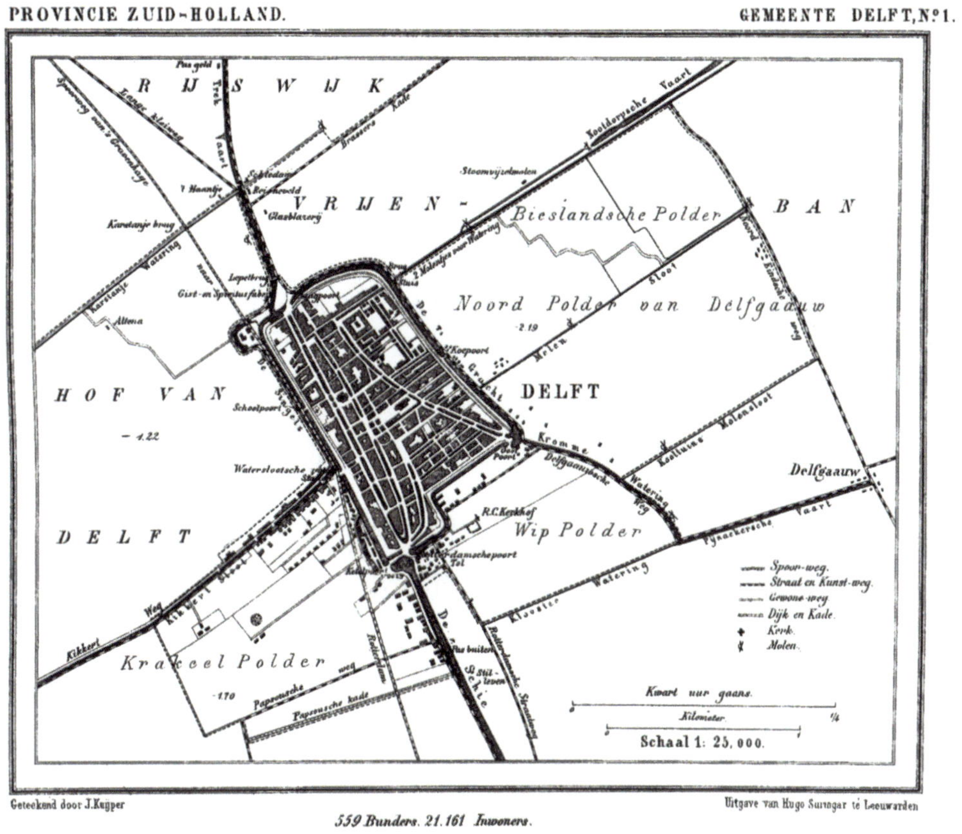

Figure 5.14 Delft around 1870.

Source: Municipality of Delft.

centre were at risk of more frequent overflowing because of more intense precipitation, a higher discharge level of the river Schie and persistent subsidence. To prevent this from happening, it must be possible during heavy rainfall to separate the canals in the east of the city centre from the outlet waterway, the Rhine-Schie Canal. Most old city centres are connected to the outlet waterway water *boezem*. They are usually situated on higher ground, which has better natural protection against water. The urban expansions, in particular in the twentieth century, were built in lower-lying polders. During heavy rainfall, the water is discharged to the outlet waterway water, and sometimes the old centre is then confronted with flooding, despite being the highest part of the urban area.

In 2001, a start was made on emergency measures on the most critical canals, the Vlaminggracht and the Rietveldgracht. Temporary dams sufficed for the short term; it was estimated that a permanent impoldering of the city centre would be needed within 10 to 15 years, because of sustained subsidence together with an unchanged outlet water level.

The Water Plan Steering Group resolved on 14 April 2003 to design a semi-permanent enclosing variant, which involved the near-term installation of fixed closing structures where possible, and flexible gates where desirable. This alternative solution was elaborated to produce a water management requirements specification for Delft city centre. The specification observed the constraints from the viewpoints of water quantity, water quality, management and maintenance, and took into account the various users of the water system. It must be possible to isolate the east of the city centre at eight places (see the ground plan) for each of which the closure options were defined. The result is a plan for building six – at all locations – movable water-retaining structures and a pumping station to remove excess water from the polder (i.e. the city centre). The two existing water-retaining structures were integrated into the plan.

5.4.3 Solutions and process planning

The photos on the following pages show the existing situation of the canals to be closed. Temporary measures have already been installed in two canals, the Rietveldgracht and the Vlaminggracht.

The city centre canals have other functions besides being part of the municipality of Delft's water system. The canals in which the water retaining structures are being installed are also used for tourism and leisure activities. A boat takes passengers on a

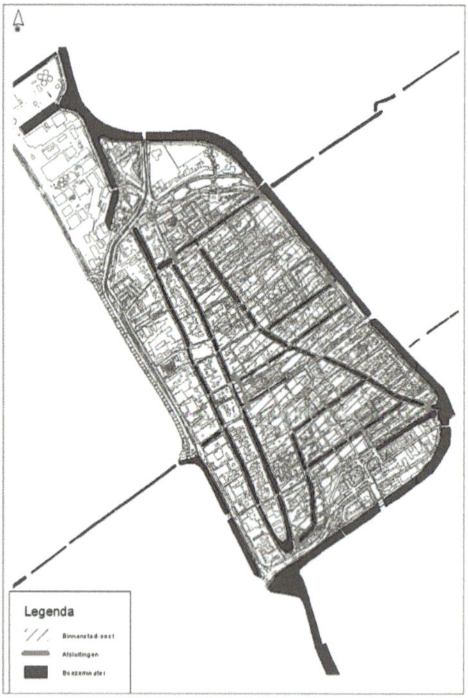

Figure 5.15.1 Proposed city centre enclosing structures.

Source: Municipality of Delft.

round trip of the Old Delft and New Delft, and pedal boats and canoes are also for hire in the city centre. Some canals (Koornmarkt, Old Delft (three boats), Binnenwatersloot, Mijnhaven, corner of Voldersgracht and Vrouwenrecht, Vrouw Juttenland) also have floating bars. The floating bars are brought into the city through the Noordkolk. The rubbish collection barge usually gains access through the Oostpoort. Preserving these water functions is a constraint in the design of the water retaining structures.

After a lengthy process, in which the urban integration of the water retaining structures received considerable attention, the municipality and the Water Board proposed tilting weir constructions for all locations. This was desirable for water management in some locations, or to maintain a free passage for floating bars, the rubbish collection barge, canoes and pedal boats. The measure was proposed for the Voldersgracht partly at the request of the urban aesthetics committee, to ensure free passage for canoes and pedal boats here too. The tilting weirs at the Oostpoort and the Noordkolk were made extra wide to maintain the link with the Rhine-Schie Canal now and in the future, allowing the floating bars and rubbish collection barge to access the east of the city. A lock construction will be built in due course prior to permanent closure in at least one of these two locations. This can be achieved with a second tilting weir and a small pumping station. The series of photos below gives an impression of the water retaining structures to be built at the locations mentioned.

Figure 5.15.2 Distribution of the areas.

Source: Municipality of Delft.

Figures 5.16.1, 5.16.2, 5.16.3 Rietveldgracht and Vlaminggracht (with temporary closures). Kantoorgracht.

Figure 5.16.4 Voldersgracht.

The water management requirements specification determines the necessary pumping capacity. A pumping station of at least 5000 cubic metres per hour is desirable in the temporary closure situation in the north of the city centre, assuming the current dominant flow direction. The pumping capacity must be increased to 8500 cubic metres per hour in

Figures 5.16.5, 5.16.6 Oude Langendijksgracht and Achterom.

Figure 5.16.7 Oostpoort. Existing situation of water retaining structures in canals.

Figure 5.16.8 Kantoorgracht. The construction to be built at this location is a tilting weir. The construction allows small pleasure craft to pass and provides sufficient through flow.

Figure 5.16.9 Voldersgracht. The construction to be built at this location is also, partly for recreational reasons, a tilting weir.

Figure 5.16.10 Oude Langendijksgracht. With regard for uniformity, a fixed tilting weir construction will be built here.

Figure 5.16.11 Achterom. The construction to be built at this location is a tilting weir. The construction allows small pleasure craft to pass and provides sufficient through flow.

Figure 5.16.12 Oosteinde/Oostpoort. A larger tilting weir will be built near the Oostpoort, which can be augmented later with a second element to allow lockage, if the need arises.

Source Fig. 5.16.1–5.16.7: Municipality of Delft, memorandum on water retaining structures in canals in the east of the city centre 2004.

the event of permanent impoldering. The preferred location is the Kantoorgracht. A combination is also possible with a second pumping station near the Kolk.

The water retaining structures were proposed in consultation with the Water Board for Delfland. It was agreed, in line with the water plan, that the municipality would act as coordinator in the construction and the Higher Water Board as the driver.

The plan below was drawn up for the design, planning and construction of the water retaining structures and the pumping station. The municipality has now completed the planning of the major maintenance to the Duyvelgats Bridge near the Kantoorgracht. The construction, a tilting weir, and the pumping station were included in the plans. Work on a construction (a stop-log recess) at the Oostpoort started in late 2004. The work was completed in early 2005. The preliminary design of the other locations started in 2005. The water retaining structures were presented to the urban aesthetics committee on 8 September 2004. The planned constructions and the chosen materialisation were well received. Additional details of the solutions worked out for each subproject were submitted to the committee.

5.4.4 What does it cost and who pays?

The scale and the comprehensive nature of solving the city centre flooding problems call for careful attention. The municipality of Delft is coordinating the project. The Water Board for Delfland is thoroughly involved and is providing expertise and supervision. The costs are being shared between Delft and the Water Board for Delfland. The Water Board for Delfland is paying for investigations and execution of the surface water part.

The Delft Water Plan makes firm proposals for implementing emergency measures at the Rietveld and the Vlamingstraat. When the Water Plan was adopted there was no insight into the long term solution. The Water Plan includes as a guide a figure of approximately 1.13 million euros to be charged to the Water Board for improving surface water management. This amount was embedded in the Water Board's ABC plan

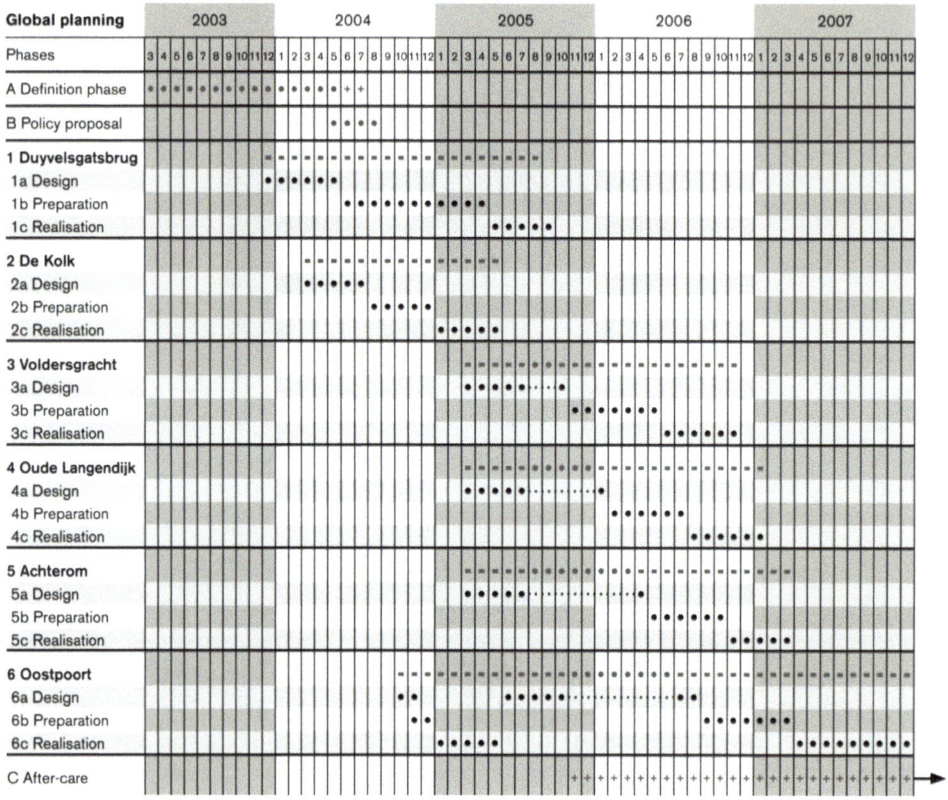

Figure 5.17 Overall planning.

Source: Municipality of Delft.

in September 2000. The municipality has earmarked 453,780 euros for the municipal part of enclosing the city centre from the EZH funds (which are the proceeds of the sale by the municipality of the shares in EZH, the South Holland power company). Approximately 195,000 euros of this amount was expended in 2004 for the emergency measures in the Rietveldgracht and Vlaminggracht, and for investigation (the future survey) and support from the urban design and architectural bureau Rappange. The amount remaining is approximately 250,000 euros. Agreements have been made with Delfland on the division of costs on the assumption that the Water Board is responsible for the functional part and the municipality for the urban integration.

The municipality of Delft, also on behalf of the Water Board for Delfland, applied for a grant on 1 April 2004 within the framework of the National Administrative Agreement on Water (NBW) under the ‘Provisional scheme for a nonrecurring payment for combating regional flooding’. The implementation plan submitted covers two projects, the city centre project and the Hap/Lap project. The grant applications for both projects were accepted. The contribution for the city centre is approximately 25% of the execution costs with a maximum of approximately 750,000 euros.

Part	Responsibility / task: Water management	Responsibility / task: Recreation	Responsibility / task: Urban design	Total costs	Costs Delftland	Costs Delft
Fixed dam	•		•	€ 50.000	€ 50.000	€ 0
Overturn dam Voldersgracht	•	•	•	€ 75.000	€ 56.250	€ 18.750
Overturn dam De Kolk	•	•	•	€ 75.000	€ 56.250	€ 18.750
Overturn dam Achterom	•	•	•	€ 75.000	€ 56.250	€ 18.750
Overturn dam Kantoorgracht	•	•	•	€ 75.000	€ 56.250	€ 18.750
Single dam Oostpoort	•	•	•	€ 500.000	€ 450.000	€ 150.000
Existing dam Rietveld	•	•	•	€ 0	€ 0	€ 0
Existing dam Vlamingstraat	•	•	•	€ 0	€ 0	€ 0
Pumping station Kantoorgracht	•1)			€ 1.000.000	€ 750.000	€ 250.000
Control and automation	•			€ 200.000	€ 200.000	€ 0
			Sub-total	€ 2.150.000	€ 1.675.000	€ 475.000
			NCW subsidy	541.306	€ 421.715	€ 119.591
			Total	**€ 1.608.695**	**€ 1.253.285**	**€ 355.409**

1) the costs for the water constructive measures are being divided between Delft and water board Delftland whereby extra pumping capacity.

Figure 5.18 Summary of provisional estimates for water retaining structures and pumping station (+/− 40%).

Source: Municipality of Delft.

Estimation	Pumping station 700 m3/hour	Pumping station 5000 m3/hour
Quay walls (50 % costs bridge-50%costs pumping station)	€ 9.475	€ 9.475
Pumping station civil and machine engineering	€ 125.030	€ 155.650
Remaining (50 % costs bridge-50%costs pumping station)	€ 27.500	€ 27.500
Electro technical	€ 108.000	€ 140.000
Duck weed fence and cleaner	€ 10.000	€ 70.000
Automate	€ 90.000	€ 90.000
Adjustment control building	€ 50.000	€ 50.000
Building costs exclusive bridge	€ 420.000	€ 542.625
Added costs = 83% (including VAT)	€ 348.604	€ 450.379
Total rating exclusive bridge	**€ 768.609**	**€ 993.004**
Difference		€ 224.395

Figure 5.19 Cost estimate of pumping plants.

Source: Municipality of Delft.

The pumping capacity needed in the city centre is considerably higher (by a factor of 3.5) than what would normally be installed for a comparable area. It was agreed that the municipality would accept the additional costs and will investigate how the additional pumping capacity could be compensated in the future. The financial comparison with two pumping plants of different capacities is shown in the Figure 5.19.

5.5 CONCLUSION

The recovery of the *singel* structure in Utrecht contributes only modestly to the issue of insufficient water storage. It does so only where an incidental separate system can be created and storm water can be discharged through that system into the surface water. However, the municipality foresees the installation of a separate system in the city centre in the long term. When this happens, there will be a great need for more storage capacity, among other things in the form of more surface water. In this sense, restoration of the *singel* structure is a necessary long-term investment in urban water management.

It is logical to use underground water storage to reduce the frequency of overflows in the Utrecht situation, because the facilities already exist. However, where this is not the case, which will be true in many other cities, it would be worth considering creating underground water storage directly adjacent to the open water, *singel*, or canal being restored.

The Breda project involves both public and private interests. It was therefore only natural to place the project development and implementation with a public-private partnership (PPP). In that respect, agreements could have been made on a distribution of the costs in proportion with the revenues. In that case, it would not have been the municipality, the Chamber of Commerce and the entrepreneurs who acted as the client for the study into the economic impacts, but the PPP, and then the outcomes of the study conducted by the BRO might not have revered the maximum variant as emphatically as it did. However, Breda did not opt for a proportional risk-bearing participation involving the municipality and private parties. Instead, a sharp separation of responsibilities was chosen. The municipality has taken responsibility for the restructuring of the public area with the plan for the 'New Mark' as the projected result.

The attractiveness of the public space was the prime consideration in the restoration of the harbour in Breda. Improving the water management in the city centre was not among the key objectives of the plans. Also with hindsight, the plans made only a modest contribution on that point. The main contribution was a possibly faster discharge of the water from south to north through the central area, which is needed in heavy rainfall when the natural watercourses bring in much water from the south. Flooding sometimes occurs under these conditions in the urban area to the south of the old centre. The plans also involve a modest increase in the quantity of surface water and in the water storage capacity in the central area. This additional storage capacity can scarcely be put to effective use in the existing situation because of the absence of a separate sewer system, in common with almost all old town centres.

As mentioned, old city centres are nearly always built on the higher locations in the landscape, which provide natural protection against flooding. However, the Delft situation indicates clearly that flooding can also happen in city centres. Indeed, the urban expansions were built in the lower-lying polders, which discharge their excess water into the outlet waterway water. The outlet waterway water is by definition the open water with the highest level relative to the surroundings, and the old centre is connected to that outlet waterway water. The level of the outlet waterway water rises during heavy rainfall because it is the main discharge for the entire urban area, and is effectively the aorta of the urban water system. The solution to this problem that has been devised in Delft can also be applied in other city centres subject to flooding.

Chapter 6

More water in the city, from 1850 to 1945: consolidation

Wout van der TOORN VRIJTHOFF and Erwin HEURKENS

Department of Real Estate and Project Management, Faculty of Architecture, Delft University of Technology, Delft, The Netherlands

6.1 INTRODUCTION

The Dutch economy went into decline after the Golden Age (seventeenth century), and building production accordingly fell sharply. Industrialisation started later in the Netherlands than in the surrounding countries, and it was consequently the second half of the nineteenth century before large-scale Dutch city expansion resumed. The districts from the period up to World War II had diverse characteristics.

At one end of the scale were the densely built districts from before the Housing Act (1850 until 1901, the first ring) and from after that time (1901 until 1918, the second ring). These districts are usually without surface water, and the nature of the subsoil determines which water issues are the most urgent. Many of these districts, such as in De Pijp in Amsterdam, were built around the agricultural polder land parcelling system. Urban renewal often improved the quality of homes in these districts, while the same cannot always be said of the spatial quality.

At the other extreme, spacious, select residential neighbourhoods and urban districts with an imposing street plan were built in the later period (1918 to 1945, the third ring). The original landscape then tended to be ignored. On the other hand, districts of this kind often have more surface water, either in the form of winding *singels* or austere canals. The housing quality is usually still satisfactory, and the most important issue is therefore preservation (consolidation). Nonetheless, water issues may still exist in these districts, as in Rotterdam's Museumpark and The Hague's Vogelwijk.

6.2 MUSEUMPARK, ROTTERDAM

6.2.1 The area

Museumpark is situated to the west of Rotterdam city centre, in the middle of the first ring. The park has a rich history, mainly in the twentieth century. The land was part of a large estate that belonged to the wealthy ship-owning family Hoboken from the sixteenth century until 1924. The municipality purchased the estate and turned it into a park landscape, in accordance with a redevelopment plan by city architect Witteveen. Some examples of important buildings in the area are Villa Dijkzight (1851), Museum Boijmans Van Beuningen (1928) and several custom-built residences in the International Style. Additions after World War II include the Erasmus Medical Centre (MC), the Dutch Architecture Institute, the Kunsthal, the Chabot Museum and the Natuurmuseum.

Figure 6.1 Rotterdam.

Source: Topografische Dienst Kadaster, Emmen.

The park was redeveloped in 1989. One of the objectives of Yves Brunier's and Petra Blaisse's redevelopment plan was to create a connecting zone between the museums and art institutions on the north and south fringes of the park. The Museumpark event grounds are now the venue for various large-scale festivities, such as the Parade, the Zomerpodium and the open-air cinema. The Municipal Executive made funds available in 1998 for implementing the Cultural Capital outdoor programme as part of the Attractive City action plan. The Museumpark was one of the projects involved. Museumpark is now part of the Rotterdam Art Axis, which also includes Witte de Withstraat and Museumhaven.

An important element of the park's urban pattern is that it is enclosed within major roads from the post-war development phase. The area is bounded to the north by

Figure 6.2 Museumpark.

Source: Municipality of Rotterdam.

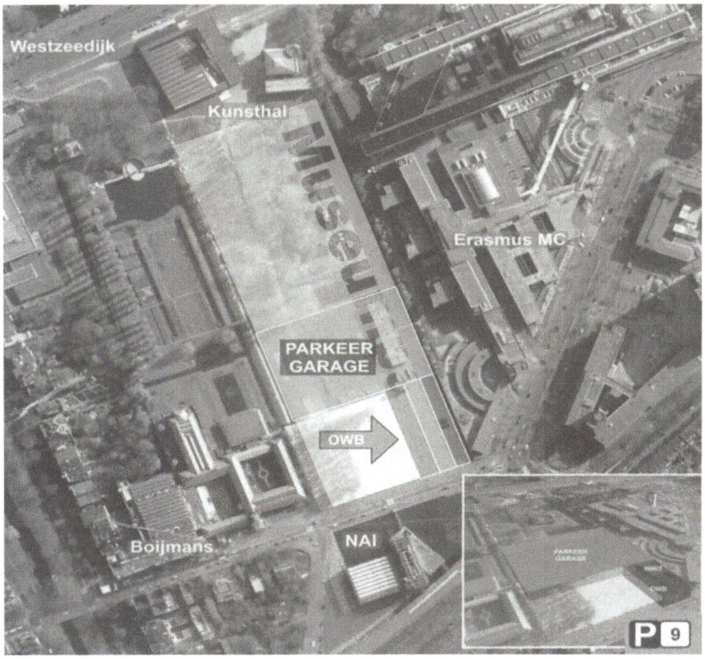

Figure 6.3 Aerial photograph of Museumpark (with north downwards) (see also color plate 15).

Source: Municipality of Rotterdam.

Rochussenstraat, which is an extension of Westblaak, an important central link crossing the city from east to west. To the east is Westersingel, the Rotterdam Central Station to Veerhaven link, which is part of the Water Project, described in chapter 2. To the south is Westzeedijk, another important traffic artery. The west of Museumpark is bounded by the Erasmus Medical Centre (MC) building, a water feature and a footpath. To the west of the Erasmus MC is 's-Gravendijkwal, which runs to the Maas Tunnel. The above factors make the area easily accessible. Museumpark is moreover a green public space surrounded by various cultural facilities in central Rotterdam.

The construction of an underground multi-storey car park, combined with underground water storage, started in the spring of 2005. The car park was built to provide approximately 900 parking spaces for visitors and personnel of the Erasmus MC. Local residents and visitors will also be able to use the parking facilities. The new car park is part of the Erasmus MC new construction project, which will largely redevelop the university medical centre over the next fifteen years. The relocation of the Erasmus MC Daniel den Hoed Centre from Rotterdam South to the centre will achieve an important bundling of training hospital facilities. The new construction will span multiple phases and years, in accordance with what is known as the 'roll-up scenario'. All care-specific functions will be situated in a single large complex, where patients' recovery will be stimulated by the creation of a 'healing environment'. An important point is the link with the city, which the design of the public space in the medical centre is intended to achieve.

Figure 6.4 Artist's impression of the Erasmus MC in 2020.

Source: Municipality of Rotterdam.

6.2.2 The water issue within the urban water plan

Rotterdam's sewers and surface water are scarcely able to cope with peak storms in the current situation. A mixture of storm water and wastewater runs into cellars and basements, and groundwater flooding is becoming more common. Furthermore, the ground level is gradually subsiding in large parts of Rotterdam. The water system has hardly any more room for contingencies.

At the 2005 International Architecture Biennial, the theme of which was 'The Flood', the municipality of Rotterdam, the Hollandse Delta District Water Board, the Water Board for Schieland and Krimpenerwaard and numerous specialists from a wide range of fields discussed how to tackle the water issue within the Rotterdam 'Diamond' (the area bounded by a system of motorways). The fact is that the water issue is set to become a very relevant topic in the existing city.

A package of measures (the Water Action Programme) for producing a sustainable water system is included in the 2000–2005 Water Plan. This is a joint product of the municipality of Rotterdam, the Delfland Water Board, the Water Board for Schieland and the Krimpenerwaard, the IJsselmonde District Water Board, the Hollandse Eilanden en Waarden Water Purification Board and the Directorate-General for Public Works and Water Management for South Holland. The management of quality and quantity must be able to provide for all the relevant designated uses. A distinction will be made between a basic package and area-specific measures. The main basic measures will be concerned with the construction, management and maintenance of the sewer system (increased sewerage charges, dredging and eliminating the maintenance backlog) and regulating the groundwater level. The sewer serves to receive wastewater, but is also suitable for removing excess water in the event of flooding. Work to connect various premises to the sewers will therefore be accompanied by expanding the sewer capacity, mainly by improving the through flow to the waste treatment plants.

The area-specific approach targets three regions: the city centre and the old urban districts (the 'Compact City'), the districts around the city centre (the 'Blue Ring') and the areas on the city fringe (the 'Green Wedges'). The Compact City activities are oriented mainly to improving water quality, embellishing water features and preventing flooding.

Museumpark is situated in the Compact City, where serious flooding occurred in 2001 following a cloudburst. A Rotterdam municipal public works department survey

Figure 6.5 Flooding of *singels* (see also color plate 16).

Source: Municipality of Rotterdam.

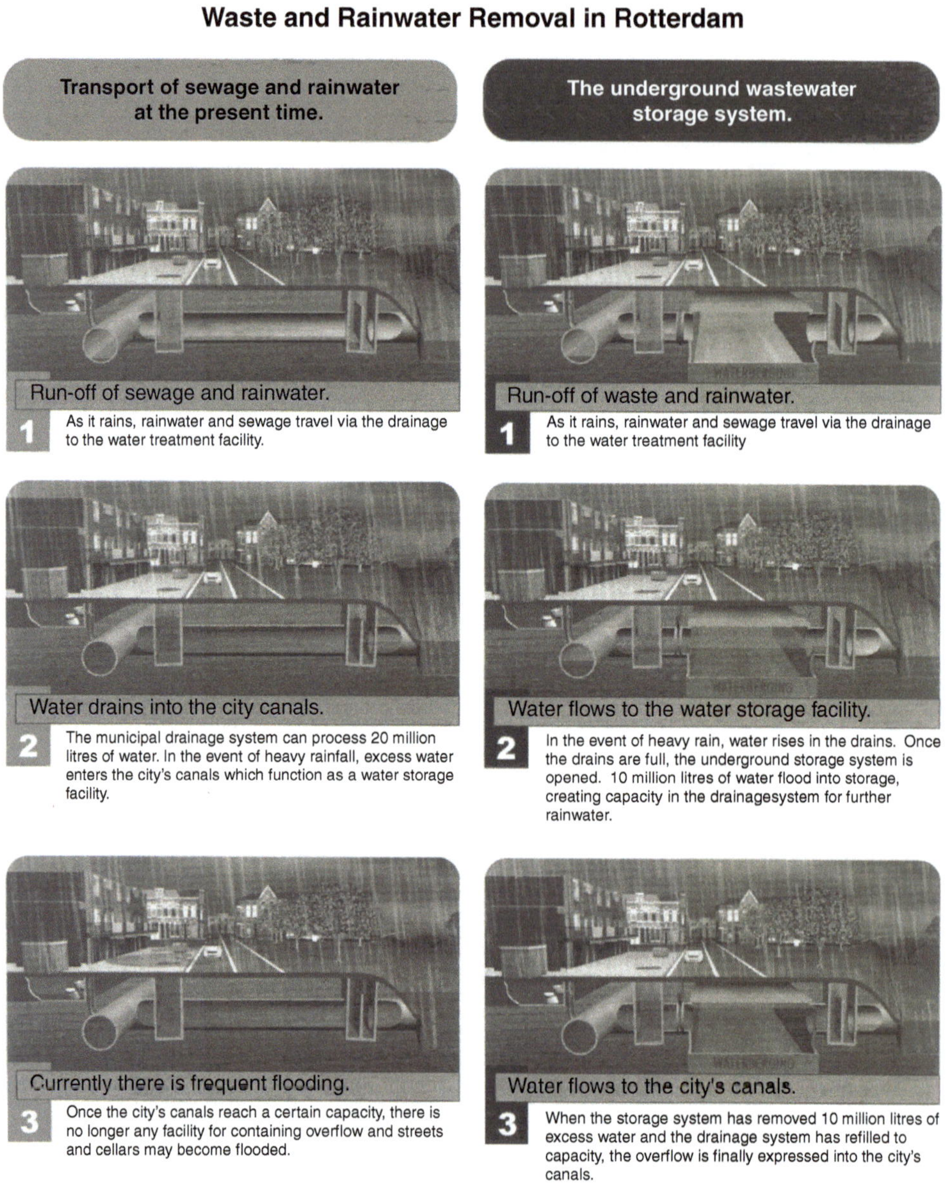

Figure 6.6 Functioning of waste and storm water drains in Rotterdam, now (left) versus the future (right). The current situation provides no overflow facilities of the sewer and in the case of an extreme rainstorm it will discharge on open water. In the future an underground reservoir will prevent this from happening.

Source: Municipality of Rotterdam.

has revealed that Rotterdam city centre has an immediate need for 40,000 litres of additional water storage capacity to minimise flooding. However, there is hardly any space for water in Rotterdam's densely built-up city centre. Innovative and inspiring solutions are therefore needed.

6.2.3 Solutions and process planning

The two urgent issues for the centre of Rotterdam are space for parking and rainwater storage. The municipality has achieved a highly satisfactory solution on both counts by building a combined underground water storage facility and multi-storey car park. However, the innovative use of the underground space in a central urban area does demand custom work. For instance, the municipal public works department and development corporation worked closely together in initiating comprehensive plans. The municipality recognises the importance of considering a broad scope of factors to the success of development or redevelopment within a complex city centre environment.

A joint answer was provided to the parking and water issues in the Museumpark project. This solution will ultimately provide one-quarter of the necessary water storage capacity in the city centre. For the project, water storage is an integral part of the Museumpark and Erasmus Medical Centre urban plan. Many parties were involved in the design and will be involved in the implementation of the multi-storey car park (see Figure 6.7).

The design process of the underground multi-storey car park-cum-water storage facility has benefited from two innovations. The first is the use of the space under the multi-storey car park ramp for a 10,000 cubic metre underground water storage facility. The ramp runs from ground level to floor −2 of the three-floor multi-storey car park. A water-retaining wall far below the ramp is needed to achieve the desired levels. This creates a closed construction pit around the ramp, where, at a modest additional cost, more soil can be excavated. The concrete underground reservoir beneath the ramp will hold up to 10,000 cubic metres of water during heavy rainfall.

	Principal parking garage	**Under groundwater storage**
	Development Company Rotterdam	Public Works, department water management
Designer	Public Works Rotterdam Architectural office Paul de Ruiter	
Executive	Public Works Rotterdam	
Contractor	Strukton (building construction)	
Installations	Consulting TS&A	
Elevator technology	Consulting Public Works Rotterdam	
Users	Municipality Rotterdam, inhabitants and visitors Centre	
	Erasmus Medical Centre	Museums around the area

Figure 6.7 Museumpark project partners.

Source: Municipality of Rotterdam.

The second innovation is to address the apparent restrictions inherent in the overground infrastructure. This formed a particularly important part of the design process in the feasibility study. Creating opportunities for the underground use of space implies abandoning two-dimensional thinking and breaking through the, at least apparent, ground level barrier to underground building. This barrier manifests itself in Museumpark in such things as the presence of a *singel*, which occupies considerable space. However, the *singel* is a shallow one, the soil mass is interrupted a few metres below ground level, and building is perfectly feasible. The use of the subsurface has made it possible to make a visual link with the Erasmus MC area in the form of a glass panelled square above the garage.

The simplest and least expensive way of increasing water storage in an existing urban area is to add surface water, for example by enlarging existing watercourses and replacing existing green strips by *singels*. Methods of this kind are often not an option in the 'first ring' and other densely populated areas, because of the limited scale of the public space, which furthermore usually has a clear purpose, as well as the little greenery available for replacement with water in these areas. The solution must then be sought in double land use. Underground water storage, if possible in combination with a multi-storey car park,

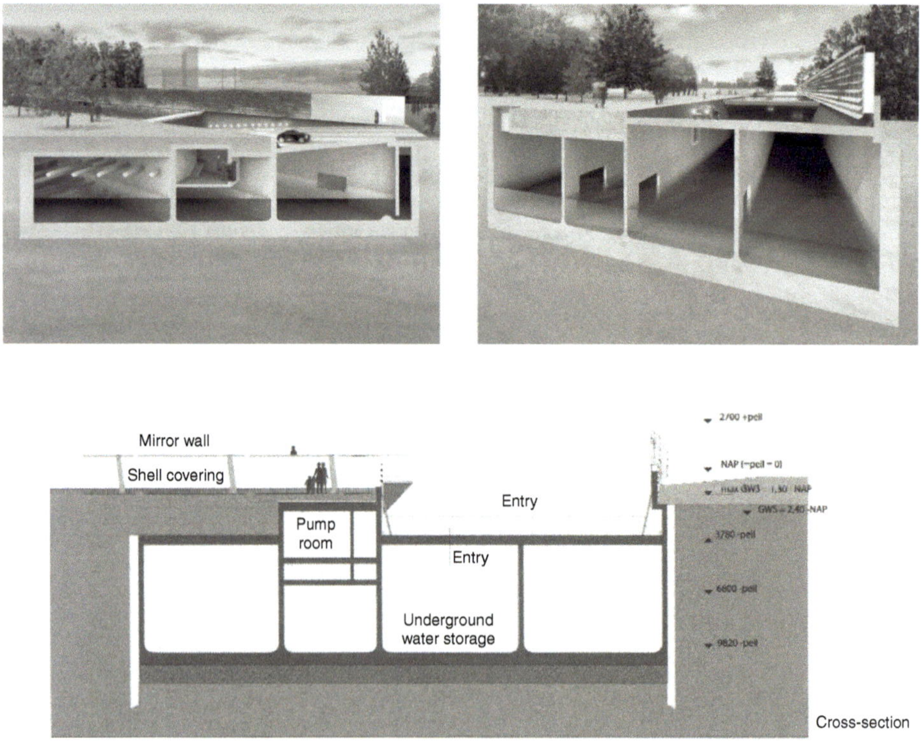

Figures 6.8.1, 6.8.2, 6.8.3 Artist's impression and cross section of the reservoir (see also color plates 17,18).

Source: Municipality of Rotterdam.

is then an option. Another is to lower parts of the urban infrastructure and to use it as water storage in periods of heavy rainfall. Squares are possibly suitable for this solution.

An additional current issue in Rotterdam is that the *singels* become polluted by the overflows from the existing combined sewer system during heavy rainfall. The underground reservoir acquires a combined function by making it an extension of the sewer system. It both prevents flooding and improves water quality. Besides the positioning as a sewer system extension, the position relative to the main pumping station is crucial to discharging water from the reservoir. The main pumping station concerned is to the south of the plan area to which the sewer system is connected. The stored rainwater from the

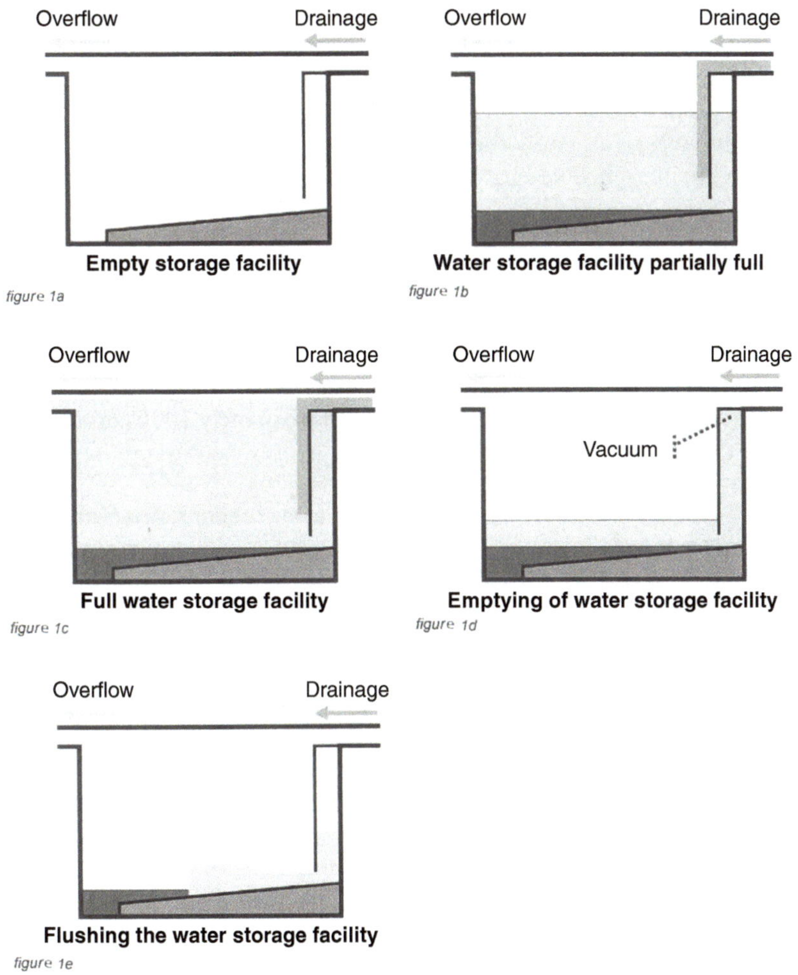

Figure 6.9 The process of filling, emptying and flushing the underground reservoir. The left arrow is overtopping towards the open water system and the right arrow is the supply from the sewer. The first picture (a) shows the empty condition, the second (b) partly filled. The third (c) is full and the fourth (d) it is emptied by vacuum. The last is the reservoir flushed.

Source: Municipality of Rotterdam.

underground reservoir is discharged to the pumping station when the weather improves, and then flows into the River Maas. Both measures have been included as targets in an area-specific urban centre approach in the Water Action Programme. The operation of the water storage system on a district level is shown in figure 6.9.

The size of the reservoir suggests that its most important function is storing water during heavy rainfall. Nonetheless, the reservoir is also designed to allow the sedimentation of sludge from the sewer (1a and 1b). If the entire reservoir capacity is used in extreme rainfall, the water then overflowing is already significantly cleaner than the current, direct overflow (1c). As a result, Rotterdam city centre will not only suffer less flooding, but also fewer bad smells and less surface water pollution. After use, the underground water storage is pumped out, during which some of the water is retained by means of vacuum (1d). The vacuum fills of its own accord when the reservoir is almost empty, and the water surges over the reservoir bottom, taking the sediment with it (1e). The remainder collects in a gutter and is pumped back into the sewer system. An underground panorama room is provided for monitoring the process of filling, emptying and flushing the reservoir.

The multi-storey car park, the underground water storage and the Museumpark redevelopment are scheduled for completion in May 2007.

6.2.4 What does it cost and who pays?

The costs associated with the least expensive way of increasing the water storage capacity, by adding surface water, are approximately 350 euros per cubic metre (when the land is made available 'for nothing' by the municipality). Achieving 1 cubic metre of storage capacity thus demands 3–4 square metres of surface water. The cost of building a single underground water storage facility is approximately 1000 euros per cubic metre. However, the costs in combination with a multi-storey car park are approximately 600 euros per cubic metre.

The total costs for the underground water storage facility are 6 million euros. The original intention was for one-third to be provided by the national government, one-third by the municipality and one-third by the Water Board for Schieland and the Krimpener Waard. However, the national government contribution evaporated at an early planning stage. The municipality is now providing two-thirds of the costs.

With hindsight, this situation is not entirely reasonable. The costs of the alternative to 10,000 cubic metres of water storage in the form of additional surface water would have been 3.5 million euros. In that case, the municipality would have made the ground available 'for nothing' and the Water Board would have borne the costs of digging the water storage facility. The reason for not following this route lies in the historical background. Not so very long ago, the Rotterdam municipal authority arranged and paid for all aspects of the water issue within the built-up area. Now, tasks, and therefore also costs, are gradually being transferred to the Water Board.

6.3 VOGELWIJK, THE HAGUE

6.3.1 The area

Vogelwijk is situated in the northwest of The Hague. It is a select residential area from the interwar years, based on the garden town concept. The area is bounded by Nieboerweg, Laan van Poot, part of the Westduin Park and the De Savornin Lohmanlaan, Sportlaan

Figure 6.10 The Hague.

Source: Topografische Dienst Kadaster, Emmen.

and Houtrustweg. Part of the Westduinen, Bosjes van Poot and Houtrust Sport Park are within the Vogelwijk boundaries. Most of the residential area is in the former Segbroek polder, where seventeenth-century farms lined the edge.

Barracks in the Segbroek polder between the current Nieboerweg and De Savornin Lohmanlaan were part of Berlage's expansion plan of 1908, to replace the Frederiks barracks and Alexander barracks. However, these plans did not materialise. Shortly afterwards, the municipality started buying up land in the Segbroek polder for a new residential area. The Houtrust Bridge over the Afvoerkanaal (discharge channel) was completed in 1909, thus providing access to the future district from the Staten Quarter. Construction of the Houtrust Sport Park took place around 1910, and the municipality built several streets during World War I. Several existing roads were then paved and named: Houtrustweg, Laan van Poot and Sportlaan. The Cooperative Housing Association Tuinstadwijk Houtrust was founded in 1917, by the architects W. van Boven and Z. Hoek, and others. The municipality stipulated in the ground lease that the land was to be used exclusively 'for building that would give the district the character of a garden town' (Lankamp and Valentijn 1992). The first stone of the residences on Leeuwerikplein and Houtrustlaan was laid in the summer of 1918. The street plan was designed by W. van Boven, while many of the houses were designed by Z. Hoek and W. Verschoor.

Vogelwijk currently has some 4,800 residents. The 56–64, 65–79 and 80+ age groups are overrepresented there compared with the average in The Hague. Vogelwijk is a spacious district with luxury homes from the 1920s and 1930s. Most of the homes are grouped in terraces of three to seven. There are a few flats in the district, in particular on Sportlaan. The homes are expensive, even by The Hague standards. The average income is correspondingly extremely high.

Neighbourhood shops can be found only around Mezenplein. The Red Cross Hospital has a prominent position on the edge of the district. The Houtrust complex was an imposing building with an illustrious past. It was the venue of pop concerts, sporting competitions (athletics) and other public events. The municipality decided to demolish Houtrust because of the poor state of the building. Homes and a secondary school are planned for the land that becomes available.

Houtrustweg, Sportlaan Segbroeklaan and Nieboerweg through roads go alongside and through the district. Around the edges of Vogelwijk are important green zones, such as the Bosjes van Poot and the Westduinen. More than 90% of the homes are owner occupied. The design of the public space (greenery and paving) throughout Vogelwijk has not been upgraded for some considerable time.

In comparison with the Segbroek urban district as a whole, Vogelwijk has:

- much greenery in and around the district;
- a high income per household;
- an extremely low unemployment figure;
- an extremely low truancy rate;
- many outdoor sports facilities.

The Haagse Beek is an important element in the Vogelwijk urban layout. This stream was created in the formation of the coast and dunes. Water would run off the sand ridges into intervening low-lying areas, where it would collect and flow away. The source of the Haagse Beek is in the current Schapenatje dune, where there was once a wet area close to the current Kijkduin to the northwest of the country estate Ockenburg. It flows from there to the northwest through Segbroek, the dune slacks lying between the sea and inner dunes. The Haagse Beek finally discharges into in the Scheveningse Veer. The course of the stream is shown below in a fragment of the map of Delfland produced by Floris Balthasars in 1611.

A recurring problem in the history of the Haagse Beek was how to lead the water in a continuous stream through the city. Drainage problems occurred because it flowed in the wrong direction, which is to say across the dunes from Ockenburg to the lower-lying Zorgvliet. A strategy was needed in order to level this elevated barrier. The simplest solution was to make the bed progressively deeper as it approached the inner dune, creating a slope in the opposite direction. Another possibility was to use mills

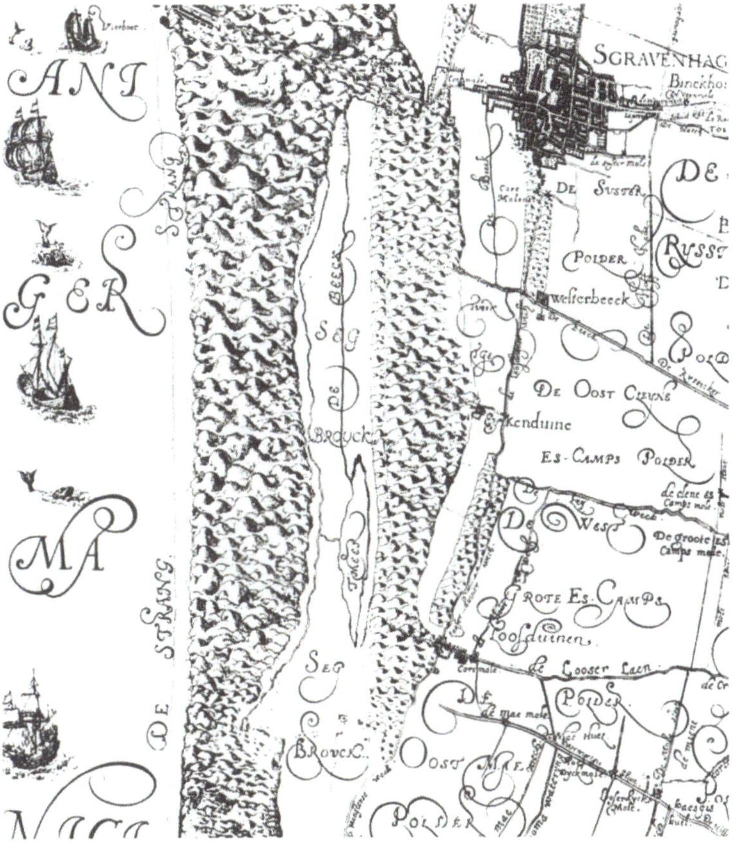

Figure 6.11 Map of Delfland from 1611.

Source: Municipality of Den Haag.

and locks to back up the water. However, the Haagse Beek's elevated location regularly caused drainage problems for agricultural land.

The Haagse Beek was tackled rigorously in 1942. A strategy employed by the German army in various west European coastal areas created what was known as the Atlantik Wall. The Hague was provided with extra defences as the seat of the German administrative agencies. The course of the Haagse Beek was used in Segbroek to supply water to the antitank trenches.

6.3.2 The water issue within the urban water plan

The book *De Haagse Beek. Een (natuur)historische verkenning* (*The Haagse Beek. A natural history survey*) (van Doorn and Mennema 1992) devotes much attention to the improvement of the water system and of the natural assets in the immediate surroundings. The Delfland Water Board adopted a new water-level decision in February 2006, among other measures for the Haagse Beek. The level was made 'flexible' in several parts

Figure 6.12 The Haagse Beek.

Source: Municipality of The Hague.

Figure 6.13 The Atlantik Wall in Segbroek.

Source: Municipality of The Hague.

of the Haagse Beek, which meant that the water level was allowed to vary between an upper and a lower limit, to the benefit of the ecology and the water quality. The surroundings consist mainly of recreational areas, such as sports grounds and footpaths, but homes and businesses are also located there. These multiple functions were taken into account in the water-level decision.

The water in the Haagse Beek is a chain of urban ponds, watercourses and substantially original stream courses. The water and the surrounding greenery form an important green-blue structure through the city. However, the Afvoerkanaal cuts the water system in two. A relatively high level is maintained in the Haagse Beek, and near Kijkduin it is at groundwater level. Water is let in from the Afvoerkanaal in order to maintain the high level, which is necessary to compensate for seepage and the discharge to the sports grounds and the *boezem*. Much of the stream is covered with loam to prevent water seepage. The incoming water from the Afvoerkanaal means that the water in the Haagse Beek is of poor quality. Overflows from the sewer are therefore no longer permissible. The Haagse Beek sub-sector therefore set improving the water quality as an objective in the 1998–2002 Water Plan for The Hague. The ambition was expressed as '*water that lives*'.

The most important objectives in this ambition are sustainable water management and the development of water-dependent natural assets. The water quality satisfies the water quality standards on all points. Any deviations are caused by local natural conditions. The surface water is no longer affected by sewer overflows or untreated discharges. The effect of diffuse pollution is limited by design and management measures. More than 10% of the banks have been laid out in a nature-friendly way, and many, sometimes rare, plants and animals are to be found there.

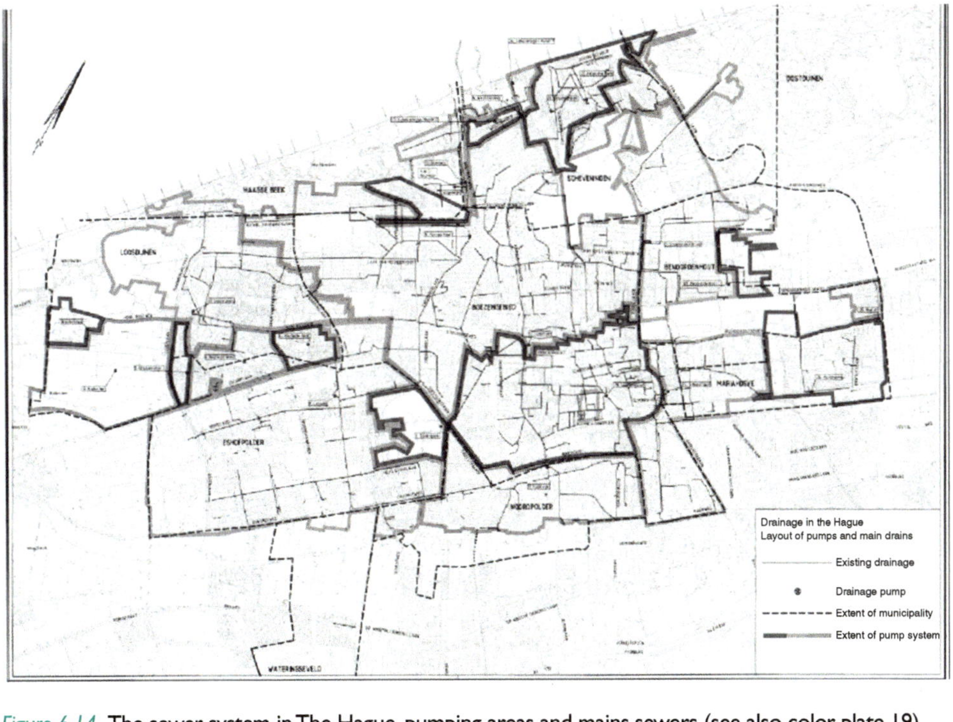

Figure 6.14 The sewer system in The Hague, pumping areas and mains sewers (see also color plate 19).

Source: Municipality of The Hague.

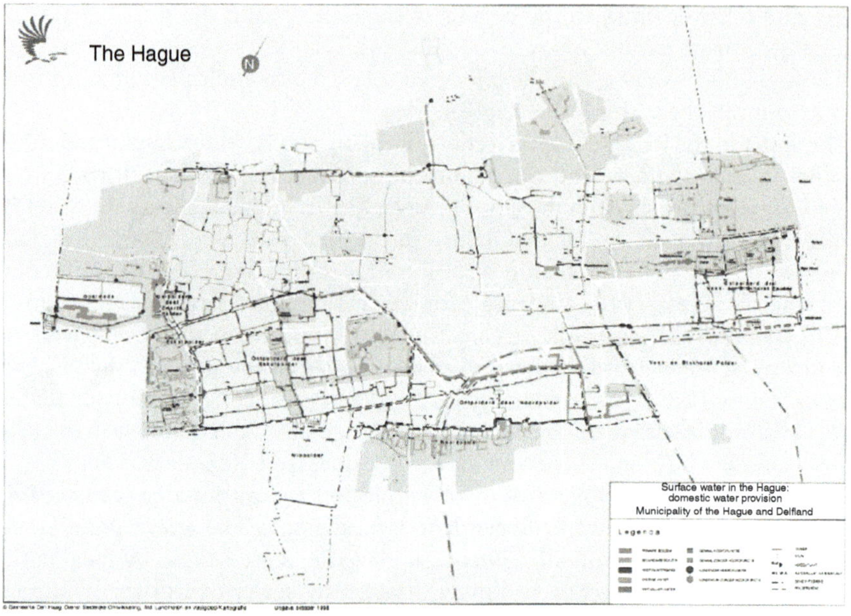

Figure 6.15 Surface water in The Hague, water management structure (see also color plate 20).

Source: Municipality of The Hague.

6.3.3 Solutions and process planning

The Haagse Beek is one of the sub-sectors in the water system of The Hague. This sub-sector has four pumping areas: Vogelwijk, Duindorp, Morsestraat and Loosduinen. A phased approach was used for the sub-sectors, in which the Scheveningen and Haagse Beek sub-sectors serve as 'pilot areas' for the approach to the others. Experience gained in the pilot areas will be used to optimise the approach in the other sub-sectors.

The Haagse Beek sub-sector executed the following sewer-related measures in the planning period to 2002, within the framework of the ambition '*water that lives*':

- tackling the sewer overflows near Noordwijkselaan and Daal- en Bergselaan; additional related facilities are necessary in the Loosduinen sub-sector (1999);
- coupling run-off from busy roads to the sewer (1999);
- replacing the sewer at Pluvierstraat and Laan van Poot (1998).

Furthermore, far-reaching measures were implemented for the surface water in the Haagse Beek sub-sector. These measures were concerned with the water management structure and the construction of nature-friendly banks. They were developed within

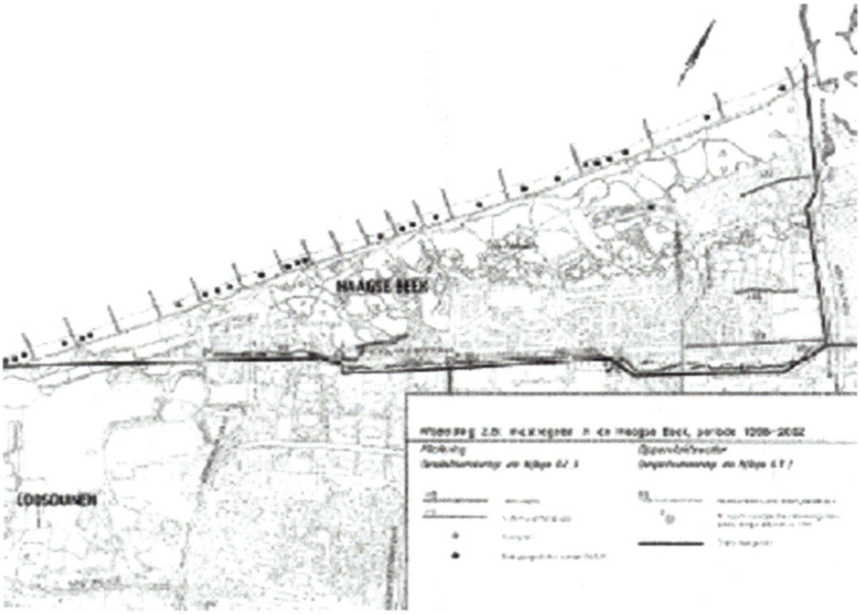

Figure 6.16 Measures for the Haagse Beek.

Source: Municipality of The Hague.

the framework of the 'Ecological recovery of the Haagse Beek and Hofvijver' project. The measures include:

- repair of the stream bed clay seal between the Afvoerkanaal and Daal- en Bergselaan to prevent seepage (1998);
- construction of nature-friendly banks and wet meadows (1998, 1999);
- modification of the water inlet of the Haagse Beek (1999);
- modification of weirs and culverts (1999);
- modification of the pumping station at the Afvoerkanaal (1999);
- construction of alternative water facilities for the Bosjes van Pex sports grounds (1999).

The Delfland Water Board evaluated the Haagse Beek project (1994 to 2003) in 2004. The evaluation revealed that several measures had yet to be executed, and that the execution of other measures was still the subject of discussion. The project comprises drawing up and executing measures as an extension of the previously executed measures within the framework of improving the water quality in the Haagse Beek. The project also provides for drawing up a management plan and completion of the transfer.

6.3.4 What does it cost and who pays?

The municipality of The Hague is responsible for the management and maintenance of the sewer system (pipes, connections and sewage pumping stations) and partly responsible for the surface water. The municipality has also taken responsibility for the shallow groundwater, to minimise potential groundwater problems. The municipal responsibility for surface water spans the quantity management of the water systems within the depoldered areas, which include parts of the Eshof and Noord polders, as well as the Veen and Binckhorst polders. The Haagse Beek also falls under the municipal management responsibility.

The Delfland Water Board is responsible for treating the wastewater and for the pumping stations that transport the wastewater from the sewer system to the Houtrust wastewater treatment plant. Delfland is also responsible for the water quality. Delfland's responsibility for water quantity management covers the part of the water system within the municipal boundaries that is not under the responsibility of the municipality.

The Water Plan for The Hague included an estimate of the municipal investment for sewer improvement, upgrade, management and maintenance, for the period 1998 to 2002. The municipality of The Hague now has a properly functioning sewer system with a long operating life, which can be maintained without high public charges. The total costs for the five-year period for sewer improvement and upgrade are estimated at approximately 55 million euros. The costs for the same period for sewer system management and maintenance are estimated at approximately 35 million euros. These costs are covered by the municipal sewerage charges, the proceeds of which are earmarked for the sewer system specifically.

The Municipal Water Tasks Act, which is part of the Comprehensive Water Act debated in parliament in the autumn of 2006, enables the investment of specific revenues for multiple water objectives. The current municipal water policy for The Hague is moreover oriented to keeping charges level. For instance, an amount is reserved for dredging work on an eight-year cycle. The municipality worked in 2006 on a Comprehensive Municipal Sewer Plan for the period 2005 to 2010. Because the sewer systems in the districts are connected in series, sewer system replacement was investigated on a district level. The costs for the Haagse Beek sewer project for the period 1998 to 2002 were estimated at approximately 1.8 million euros. All sewer overflows in the Haagse Beek area were removed.

The Water Plan for The Hague also estimates the municipal investment costs for improvement, upgrade, management and maintenance of the surface water in the period 1998 to 2002. The total costs for the five-year period for surface water improvement and upgrade are approximately 3 million euros. The costs for the same period for surface water management and maintenance are estimated at approximately 45 million euros.

6.4 CONCLUSION

The Rotterdam solution is also applicable in other densely built districts. It is not unusual to resolve the lack of space in city centres and in the first and second rings by introducing double land use. The Museumpark project demonstrates that solving at least part of the city centre parking problem can go hand in hand with implementing some of the necessary water storage capacity expansion. However, a constraint is that comprehensive interventions of this kind must be made near main sewers and main pumping stations, which demands a creative attitude to process planning on the part of municipal project leaders.

Lessons can also be learned from the organisation of this project for other locations. The various departments of the municipality of Rotterdam have been and continue to be motivated to employ a comprehensive approach and working method. This enhances the opportunities for devising and executing creative spatial solutions. This underground solution is worthy of repetition as a relatively rapidly realisable solution for the storage of peak storm water discharges. This solution is relatively expensive compared with water storage above ground. The utilisation of surface water is also better from an aesthetic viewpoint, and because of the greater visibility involved. However, the surface water solution often proves difficult in view of the current lack of space in city centres.

There are few opportunities for optimising the solution. A question worth asking in the case of the combined underground multi-storey car park and water storage in Museumpark is whether an optimum balance has been achieved between construction costs and water storage capacity. There appear to be technical reasons for not using a possibly greater underground water storage volume as part of the total project. However, the cost savings through double land use are considerable. This then facilitates a solution above ground in the form of greenery and water, doing justice to the term 'multiple use of space'.

The essence of the project in The Hague is that a separate sewer system has been created while optimising the surface water. The measures in the Haagse Beek will ultimately improve the water quality. The Vogelwijk residents probably benefit from these area-specific measures because they create more clean 'green' and 'blue' areas. Despite the considerable capital expenditure necessary, the amount appears reasonable when viewed on a per household basis. The investments can be covered from the funds formed by sewerage charges and the Water Board charges.

Chapter 7

More water in the post-war city: restructuring

Wout van der TOORN VRIJTHOFF[1] and Anita Te LINDERT[2]

[1] *Department of Real Estate and Project Management, Faculty of Architecture, Delft University of Technology, Delft, The Netherlands*
[2] *Municipality of Leidschendam-Voorburg, The Netherlands*

7.1 INTRODUCTION

The rate of building production achieved to eliminate the housing shortage in the period after World War II was enormous. Expansions were inspired by the CIAM philosophy, with much light, air and space in an open structure. The key idea was the 'neighbourhood concept', and the building pattern was of repeated 'stempels', or imprints, of high and medium rise building. At a later stage mega-structures were also built (as in Amsterdam's Bijlmermeer), and, later still, the home zone (woonerf) became fashionable. The various networks (road traffic and pedestrians as well as water) became increasingly separated. Water was then no longer seen as an integral part of the urban design, but as something separate that was often situated on the district fringes.

The post-war districts are currently undergoing large scale restructuring, and good use is being made of the large amount of public space that is typical of these districts. However, the water structure – and the desire to extend it – is sometimes under pressure. Nonetheless, the restructuring in many post-war districts is being taken as an opportunity to contribute to the water issues in the city, as can be seen in Delft, Dordrecht and Haarlem.

7.2 POPTAHOF, DELFT

7.2.1 The area

Poptahof is an expansion district from the early 1960s, designed by Prof. S.J. van Embden. Among the reasons for building the district were to provide replacement housing for the original residents of the south part of Delft city centre. This area, which is now better known as the Zuidpoort, was demolished and rigorously upgraded in the 1960s.

Poptahof is approximately 800 metres from the old centre and was built for families. The urban plan used urban design 'stempels', which were characteristic of post-war reconstruction. Eight building blocks with ten-floor gallery flats alternated with four-floor gallery flats and single-family dwellings. The district is situated within four major roads, Papsouwselaan, Martinus Nijhofflaan, Provincialeweg and Westlandseweg. Tram line 1, which links Scheveningen via The Hague with the Tanthof district of Delft, runs along two sides of Poptahof. Along the edge of Poptahof are an elderly people's complex and the In de Hoven shopping centre, where the Hoven passage is also situated. Many businesses are also located on Papsouwselaan.

Figure 7.1 Delft.

Source: Topografische Dienst Kadaster, Emmen.

Poptahof has a total of 1011 dwellings. Almost all are rented, and managed by the Woonbron housing association. The original residents have now made way for a population of almost 2800, in which more than 30 different cultures are represented. The technical and functional obsolescence and social market developments have prompted restructuring plans.

The restructuring plans already have a history of over ten years. The original expectation was that the segment of the housing stock represented by Poptahof would fall vacant. Plans were therefore developed with a focus on demolition and new building of a different type of home. However, the economic downturn and stagnation in the production of homes renewed the demand for inexpensive rented accommodation. The

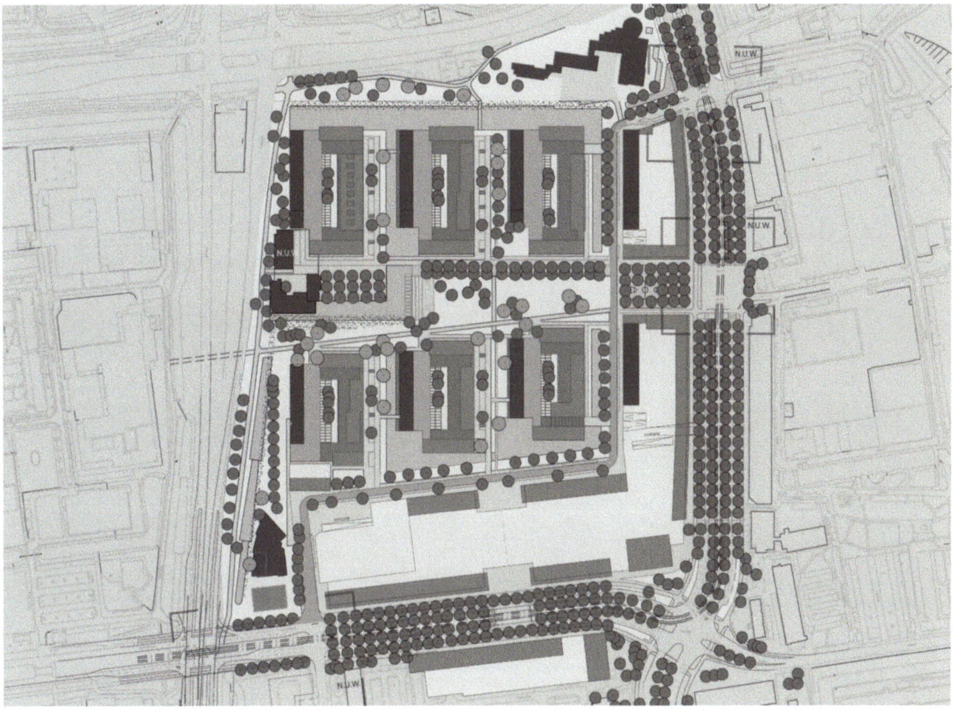

Figure 7.2 Planning map of urban design plan with suggested building fields. Poptahof was built in the sixties for 2800 inhabitants in 1019 houses, mostly stacked and 99% rental (see also color plate 21).

Source: Municipality of Delft, Van der Vorm Vastgoed and Delftwonen 2001.

Poptahof restructuring plans were amended in the light of these developments. The document *Werelds wonen in de Poptahof* (*Worldly living in Poptahof*) added details to the plans (Municipality of Delft, Van der Vorm Vastgoed and Delftwonen 2003). The document was commissioned by the municipality of Delft, the Woonbron housing association and Van der Vorm Vastgoed (the owner of most of the shopping centre and the office building, about half of which is rented to the municipality).

Most of the district, which occupies approximately 18 hectares in total, has a residential function. This part of the area consists of a northern and southern part, separated by a park-like intervening area with a water feature. There are also green strips and *singels* on the west edge alongside Provincialeweg and the north edge alongside Westlandseweg.

7.2.2 The water issue within the urban water plan

The Delft Water Plan is a joint product of the municipality of Delft and the Delfland Water Board (Municipality of Delft and Delfland Water Board 2000). The Poptahof district forms an explicit part of the Water Plan and serves as an example project. The urban design is an important link in translating the 'sustainable water management ambition' to the practice of urban renewal in Poptahof. The objective of an upgraded water system in Poptahof is to improve both the environmental and spatial quality. It will then be possible in the future to create sufficient space for storing the water during

Figure 7.3 Poptahof's characteristic spatial structure.

Source: Aedes fotoarchief.

Parts of the water plan	Fixed	Norm surface water	Standard	Increase of surface water in the area
Basic study water plan				
Water plan in a blue network	Aug. 2003	6-8% of total	325 m3/ha	Poptahof 1,80 ha
Water system analyses	Feb. 2005			
Concept water structure vision	March 2005			

Figure 7.4 Overview policy documents.

Source: Municipality of Delft.

heavy rainfall, and the water will become cleaner. Furthermore, the spatial quality that water can contribute to the living environment will be better utilised.

The water aspects of the Poptahof restructuring plans must lead to sustainable integrated water management and a self-contained area water system, which implies that less water should be let in from the surroundings. The water system will then have its own level variation, in which the water will flow through a circular system dedicated to the area. Sustainable integrated water management refers not only to the water engineering aspect, but also to the enrichment of the residential environment. This demands an improvement in quality both of the water and of the potential uses. Water buffering must also take place in the district, by storing water as surface water. The municipality's Water Plan is the basis for making water one of the most characteristic spatial elements in Poptahof.

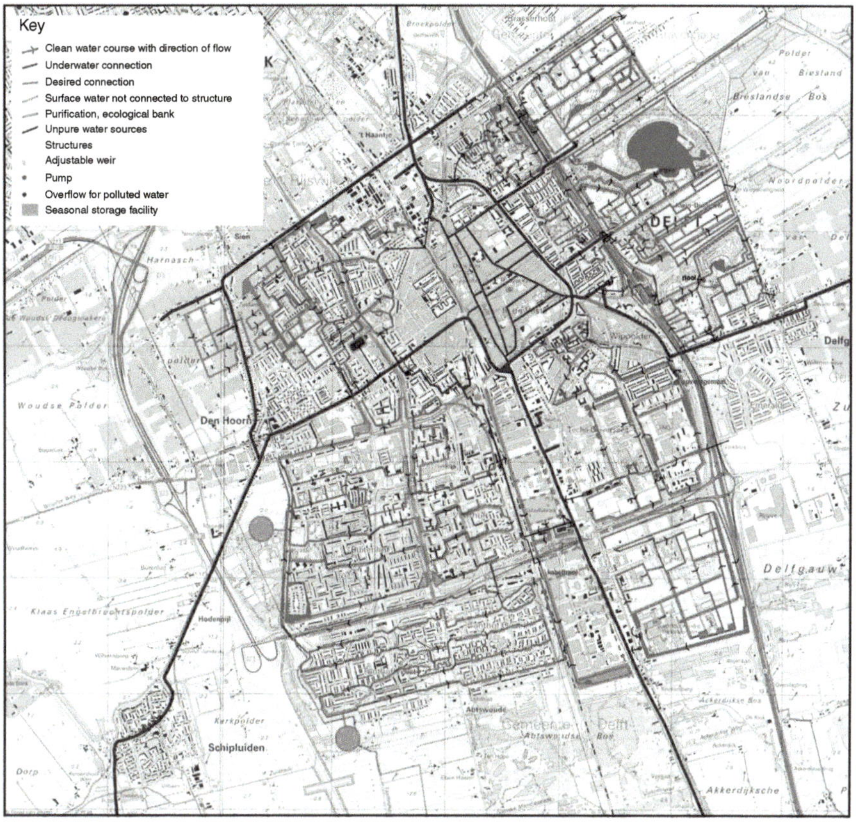

Figure 7.5 Water structure vision Delft (see also color plate 22).

Source: Municipality of Delft.

7.2.3 Solutions and process planning

Relevant administrative starting points and urban developments were considered in the Poptahof plans. Delft has run out of expansion areas for housing. The municipal boundaries have been reached, and the only remaining option is 'infill development', which will be accompanied by a certain degree of urbanisation. Poptahof's location near the city centre means that the high-rise building fits in an environment of urban concentration, and therefore has future value. The In de Hoven shopping centre and its surroundings act as a district heart for the Voorhof and Buitenhof estates.

The Delfland Water Board was an important partner in producing the Water Plan. The municipality liaised on the restructuring with the housing association and the owner of the shopping centre with a view to producing a joint master plan for the district. The coincidence of problems of a varied nature (the public space belonged to the municipality, public housing problems to the housing association and the decline of the shopping centre to Van der Vorm Vastgoed) urged the municipality to vigorously pursue the Poptahof redevelopment. Water was included in the plan in particular as an ordering principle. Water was a crucial factor from the outset. It was therefore possible to formulate a joint study question in which water had an important part. The municipality set constraints on the urban design bureau Palmboom & Van den Bout in the terms of reference for the master plan. Water had to be one of the most characteristic spatial elements in Poptahof. The following aspects were taken as starting points:

- making water, the water features and canals liveable, visible and usable for the residents;
- improving quality, both of the water and of the potential uses;

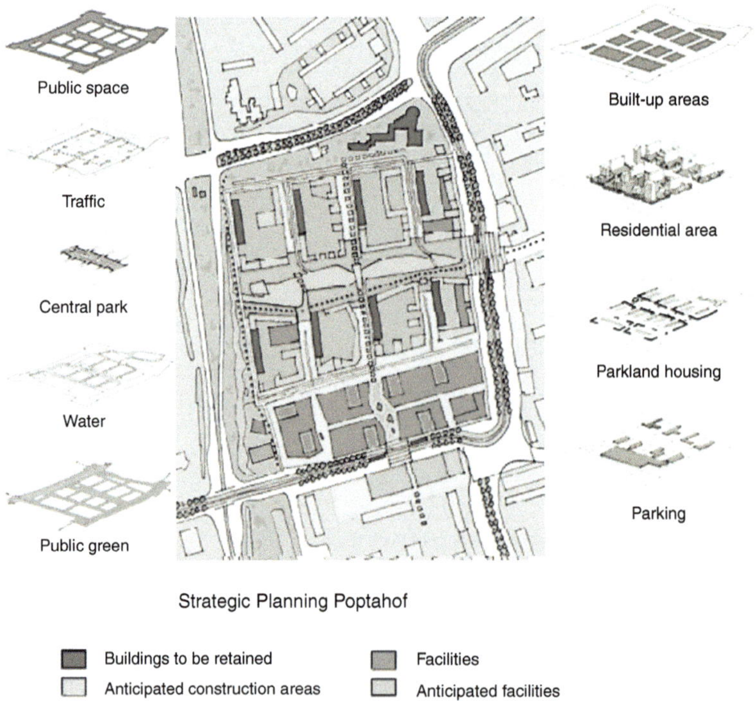

Figure 7.6 Poptahof strategic map (see also color plate 23).

The pictures on the left show (from top to bottom) the mould (public space), traffic, central park, water, public green. On the right contra mould (building foot print), dwelling landscape, park living, parking landscape. The colour legend shows in red the existing buildings, in pink what is to be build, orange the facilities and light orange the buildings near of on the facilities.

Source: Municipality of Delft.

- retaining water in the district (water buffering);
- preventing overflows.

Much effort was put into communication and participation in drafting the master plan for the district. High priority was given to allowing residents and their children to participate in redesigning the district, in order to generate public support. One of the municipality's social objectives was to maximise the consensus between the parties responsible for the water and the ultimate users. Use was accordingly made of child participation, specifically oriented to the water (Van Eijk and Tjallingii 2003). Poptahof residents represent 33 different nationalities. The children speak Dutch, and usually served as interpreters for their parents. The children were requested to interview their parents regarding water use and water awareness. A children's water afternoon in the mosque was also organised for them, where they could contribute their thoughts on themes such as water and nature, and water and games. The winning idea, a water playground, will be built. The participation of children was a great success. The parents also became involved in the water issue through the children. Insight was thus gained in a simple way into their perception and knowledge of water in the district.

The Poptahof restructuring, besides spatial interventions, also involves social and economic renewal. The ambitions for sustainability and the starting points for the intended programme are also stated in the master plan. These ambitions have been translated into the strategic map for Poptahof, which presents the future picture envisaged by the parties. This has been further elaborated into programme frameworks for the renewal and the phasing in sub-sectors. The project will be phased in order to keep the process manageable. Another advantage is that it ensures flexibility, allowing responses to be made to changes in the context, including on the housing market.

The master plan provides for increasing the surface water quantity in Poptahof, in particular by widening the *singels* at the edges of the area, while also adding new surface water in the area.

7.2.4 What does it cost and who pays?

Expansion of the scale of surface water is often the least expensive solution to the problem of a shortage of storage capacity. It was agreed in Delft that the municipality would, where necessary, provide its land 'for nothing' and encourage property developers to do the same. The Water Board is implementing the water storage. A guide figure for the implementation costs is approximately 100 euros per square metre. Surface water maintenance management used to be in the remit of the municipality. It has been agreed to transfer much of this responsibility to the Water Board.

The financial arrangements for the Poptahof are different. The costs of implementation and expansion of the water are included in the total planning costs for restructuring the district. These costs are incorporated in the land costs and will be apportioned to the property costs. This account will be paid by the Woonbron housing association and Van der Vorm Vastgoed. The Delfland Water Board is making no financial contribution to the costs of expanding the water storage in this case. The Poptahof property occupiers, like all other property occupiers, will pay water board charges, sewer charges and purification charges. Because some of the costs have been incorporated into the property prices, the property occupiers in Poptahof are paying a little more for the solution of the urban water issue.

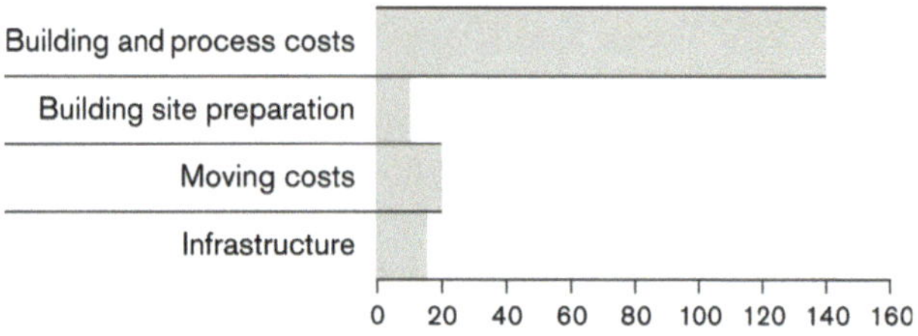

Figure 7.7 Cost distribution of Poptahof restructuring.

Source: Municipality of Delft.

The municipality estimates the costs of restructuring Poptahof at approximately 180 million euros. The costs can be categorised as infrastructure, site clearance, laying permanent roads and landscaping, removal expenses and, the largest component, construction and legal costs.

The preliminary cost estimate for a number of the operations to be performed for the Poptahof water system is 600,000 euros. These costs can be categorised as design and layout, weir construction, removal or relocation of the sewer overflows and the construction of water diversions. The total 'water costs' as a proportion of the total Poptahof restructuring costs work out at approximately 0.3%. This is approximately 7% of the costs involved in site clearance, laying permanent roads and landscaping in Poptahof, assuming that these costs are included under this heading. It should be noted that the costs for site clearance, laying permanent roads and landscaping include costs for installing sewers.

7.3 WIELWIJK, DORDRECHT

7.3.1 The area

Wielwijk is in southwest Dordrecht, adjacent to the A16. The district consists of a small pre-war part, the Zeehaven quarter, and the much larger post-war Wielwijk. Wielwijk is a typical CIAM district (built in accordance with the principles of the Congrès Internationaux d'Architecture Modern): a district with a strong urban design concept, with much light, air and space, and many facilities. A new building location was added to the district in the 1990s, the Dordtsche Hout. This part of the district consists entirely of owner-occupied housing in the middle and expensive segments and is situated to the south of the Wielwijk Park, the green area of this urban district.

Residents of Dordrecht's run-down city centre in the 1960s were keen to move to the new rented accommodation at the edge of the city. As time passed, ever more different and more modern districts appeared around Dordrecht. Wielwijk lost popularity, among other things because much of the greenery in the district had been turned into tarmac and stone to satisfy the growing demand for parking space. Little has remained of the district's good image in the last fifteen years or so. The district has even been described as the biggest porch-accessed flat estate in the Netherlands. Wielwijk comes out below the Dordrecht average on the assessment of dwellings, living environment,

Figure 7.8 Wielwijk was built at the end fifties and beginning of the sixties for 5822 inhabitants in 2855 houses, 69% stacked and 90% rental.

Source: Municipality of Dordrecht 2004.

facilities and housing quality. The district also has a higher than average throughput. Many single-parent families live there, and the proportion of welfare recipients is higher than elsewhere in Dordrecht. The district has a weak social structure, and the social cohesion is accordingly under pressure.

Performance agreements for West Dordrecht were set out in 2004 in the document *Dordrecht-West op stoom* (*West Dordrecht gets up steam*), (Municipality of Dordrecht, Stichting Interstede and Woondrecht 2004) in which the municipality and the associations agreed on steps to oppose the selective exit of higher incomes and the arrival of lower incomes, and measures to stabilise the liveability and level of facilities. The district outline *Wielwijk naar het midden en op orde* (*Wielwijk to the centre and in order*) of December 2004 set out the policy priority areas on a district level for the coming two to four years.

7.3.2 The water issue within the urban water plan

The Dordrecht Water Plan was drawn up by the municipality, the De Groote Waard District Water Board and the Hollandse Eilanden en Waarden Water Purification Board, the water partners for Dordrecht (Municipality of Dordrecht, Hollandse Eilanden en Waarden Sewage Treatment Board and De Groote Waard District Water Board

Figure 7.9 Dordrecht.

Source: Topografische Dienst Kadaster, Emmen.

2002–2003). In the Wielwijk district (and Crabbehof, which are viewed by the Water Plan as a single district) 2% of the surface is to be converted into surface water. Furthermore, a link is to be made with the Dordtsche Hout in order to dispose of the remaining excess water. Wielwijk can link the Urban Ecological Structure zones Oostkil and Dordwijkzone.

Redevelopments are being planned and implemented at various sites in Wielwijk. Opportunities for retaining and storing water are being created in these projects, in

Parts of the water plan	Fixed	Norm surface water	Standard	Increase of surface water in the area
Exploring note	June 2002	2%	10%	2,3 ha
Vision note	June 2002			
Plan of measures	June 2003			

Figure 7.10 Overview policy documents.

Source: Municipality of Dordrecht.

particular by adding water surface area. One of the projects now at an advanced stage in which water plays an important part is the Admiraalsplein shopping centre redevelopment. This redevelopment will link the north and south sides, so that the waters to the north of Wielwijk and the Dordtsche Hout will become a single system.

Singels (a Water Plan category) will blend in excellently in the building and the public space in Wielwijk. The *singels* from the adjacent Krispijn district will be extended to the Wielwijk Park via Admiraalsplein and Trompweg. Wielwijk Park marks the transition to a more open zone with watercourses (channel water category). A wet dike foot can also be created in Wielwijk along the Zuidendijk. The future increase in level in Wielwijk and Crabbehof is forecast to be 0.25 metre. Even with the redevelopment, Wielwijk will still have insufficient storage capacity. The *singels* therefore also serve as a means of water transport to the Dordtsche Hout.

7.3.3 Solutions and process planning

The implementation of the water in the Wielwijk development was initiated primarily from the concept of liveability, for which the document *Waterlint door Wielwijk* (*Water ribbon through Wielwijk*) was drafted (Municipality of Dordrecht 2001). The municipality of Dordrecht Water Plan was developed later, and the plans for the surface water (covered by the *Water ribbon through Wielwijk* document) were incorporated.

Various initiatives were taken for the participation of the residents in order to create support for the restructuring among the residents and other social organisations. New water will be introduced at a number of locations in the district where homes now stand. The homes to be demolished are going because of the redevelopment of the district, and not only to make room for water. This is also the message in the communication to the district residents. There is therefore no specific resistance to introducing surface water. Various neighbourhood evenings have been held to discuss the Water Plan and its implementation, so that residents have an opportunity to contribute their ideas. The Water Plan authors used the knowledge and the wishes expressed by the residents.

Social bodies, schools and associations were also closely involved in the restructuring of the district. The district has been designated as one of the targets of the Ministry of Housing, Spatial Planning and the Environment's 56-district approach, which stimulates urban renewal by keeping a close eye on developments and providing facilities. The involvement of the district water board will be sought for the implementation of the Water Plan in Wielwijk. At the time of writing the *Water ribbon through Wielwijk*, the district water board saw some opportunities for providing additional surface water and undertook to become involved.

An important element of *Water ribbon through Wielwijk* is a new *singel* through the plan area, which is in line with the revitalisation plan's objectives. The new and existing *singels* have an important transport function in the water system. The *singels* also have a cultural character and support the urban structure of the districts. Intersections of water with streets and roads preferably take the form of bridges, and existing culverts will be replaced where possible with open water.

Wielwijk needs 2% more surface water to satisfy the internal water requirement. A link must also be made with the Dordtsche Hout to dispose of the remaining excess water. The intention of adding water is to improve the living environment in the district and the image of the shopping centre. The municipality is convinced that introducing the watercourse is compatible with the chain reaction of changes that started

Figure 7.11 Wielwijk (and Crabbehof) water structure.

Source: Municipality of Dordrecht.

with the adoption of the restructuring plan for the district. The redevelopment is being prepared and executed at various locations. Opportunities for retaining and storing water are being created in these projects. However, even with the redevelopments, Wielwijk still has insufficient storage capacity. The *singels* therefore also serve as a means of water transport to the Dordtsche Hout. The water quality and quantity managers are involved in the planning and implementation of the process.

7.3.4 What does it cost and who pays?

Some sub-sectors of the Dordrecht water system have insufficient storage capacity, and others have excess storage capacity. The construction of several new links in the water system can allow a sub-sector with a storage surplus to compensate for insufficient storage in another sub-sector. The National Administrative Agreement on Water stipulates that the costs for the links are an expense of the De Groote Waard District Water Board.

Spatial plans to be executed in the coming years at various sites in Dordrecht include housing and restructuring. Additional water areas will be created where possible in projects of this kind. There must be 51 hectares more of open water in total in 2050 than in 2002. Approximately 14 hectares of this total is intended for receiving the water attributable to the paving of new build areas, and approximately 37 hectares for receiving water from areas already built.

Property developers are sometimes involved in restructuring projects. They will be obliged to contribute to some of the costs of the development measures, including 'water'. Space for water must be a full part of the design of the public space. It will be possible to add value to the project by stating clearly that water has to be integrated as a reinforcing element into a larger structure. 'Living on the water' is an obvious example. This benefit can be translated into a higher private contribution in the total project expenses.

Private funding of a different kind can sometimes be agreed for housing corporation restructuring projects. The water assessment and the section on water in the zoning plan constrain the water partners and ensure that both the quantity and layout of the water comply with the concept memorandum. The proportion of open water to be assumed for a zone destined for new building is 7–10% of the area, depending on the permissible rise in level in the zone during peak precipitation.

The terms of reference of the three water partners were used as a basis for distributing the costs:

- De Groote Waard District Water Board: water quantity management;
- Hollandse Eilanden en Waarden Water Purification Board (ZHEW): water quality management;
- municipality of Dordrecht: management of the sewer system and the public space, owner of the water.

The costs of each project were distributed on the basis of the distribution of tasks between the water partners: the introduction of water is paid for by the district water board, and sewer-related measures by the municipality. The necessary costs for investigating the correct measure for each project are shared between the water managers and the municipality. The municipality of Dordrecht estimated the average costs for introducing new surface water at 30 euros per square metre. However, various calculations have shown that the costs for introducing additional surface water in Wielwijk are far higher (see the calculation below).

> How much are the costs of extra surface water?
> In the period until 2050 about 37 hectare extra surface water needs to be paid out of the water plan budget. That means yearly 0,75 hectares. How much is that per year?
>
> - purchase € 5 per square meter;
> - digging € 5 per square meter;
> - transport and disposal € 10 per square meter;
> - construction € 5 per square meter;
> - contingencies € 5 per square meter;
> - TOTAL: € 30 per square meter.
>
> Per year the costs are estimated at € 225.000, or more than € 1.50 for each inhabitant of Dordrecht
> Estimates are also made for extra water surface in the areas of Dubbelmondepark and M.H. Trompweg including construction, general costs, profit and risks, contingencies and tax; purchase of land not included, removal and resituating cables and pipes and soil investigation
> The costs for Dubbelmondepark are estimated at precisely € 23 when the soils are clean and the asphalt without tar, and € 43 when the soils are polluted and the asphalt has tar. In both numbers, the costs are including detailing and furnishing and correspond with the first indication of € 30
> The costs for the M.H. Trompweg are roughly estimated and are much higher that the costs for the Dubbelmondepark. Even with a minimum scenario of realizing parking lots and the planning of trees, counting of the soil to be clean and the asphalt not to have tar, the costs will rise to € 200. This number is far from the initial calculation and shows that it is cheaper to make the water storage in an area that is green and not build up
> (Dordrecht Water Plan cost estimates for additional open water).

The distribution agreed by the district water board and the ZHEW for any costs incurred in constructing nature-friendly banks (oriented to improving the water quality) is in accordance with the *Urban water plans manual* (A publication of the South Holland South Alliance for Comprehensive Water Management, October 1999). The costs of measures for improving the sewer system are likewise distributed on the basis of this manual. This means in general that the municipality is accepting 55% of the costs and the ZHEW 45%. The ZHEW contributes only to external costs.

Various fellow public authorities have co-financing facilities for development measures, provided they are compatible with certain general or specific policy objectives, such as urban renewal, nature conservation and water storage. The measures plan will survey the opportunities and conditions for co-financing specific measures, with a particular emphasis on identifying win-win situations.

The district water board was initially extremely enthusiastic about the plans for the 'Water ribbon through Wielwijk'. It recognised that participating would enable it to implement new surface water in the future. It was also willing to invest in order to achieve their standards and objectives. However, it has now become rather more reticent, and first wants to investigate the real necessity of implementing new surface water at this site in more detail. This investigation is yet to be executed. If the district water board later changes its mind about co-financing, the funding for constructing the watercourse will have to come from other sources.

7.4 SCHALKWIJK, HAARLEM

7.4.1 The area

The area known as Schalkwijk became part of Haarlem in the early 1960s, following a revision of the municipal boundaries. Schalkwijk is a typical post-war district and was built largely in the 1960s and 1970s. The district is in the southeast of the municipality of Haarlem, between the Spaarne, the Haarlemmermeer Ringvaart and the arterial road to Schiphol. Schalkwijk is largely characterised by rectangular land parcelling, open strips and a repetition of 'stempels'. The dominant feature is 1960s high-rise building, which was created to alleviate the housing shortage. Schalkwijk's fairly brief history has produced not only a rather monotonous structure, but also a housing supply variety that is too limited for a district of its size. Schalkwijk is the largest urban district of Haarlem and covers one-quarter of the municipality's total territory.

Schalkwijk's dominant function is residential. There are many multi-storey dwellings (62%) and few owner-occupied homes (26%). The many rented dwellings are owned

Figure 7.12 Schalkwijk was built at the beginning of the fifties for 31.704 inhabitants in 14.825 houses, 62% stacked and 74% rental.

Source: Municipality of Haarlem.

by the three Haarlem corporations. This creates an opportunity for a cohesive and coordinated approach to the sub-sectors (RIGO 2000). The district also has many large office buildings, a hospital, primary and secondary schools, a professional education college, and a large shopping centre. These facilities jointly provide 15,000 jobs. There is a great deal of public space in the district, but it is situated mainly along the edges, with almost no differentiation.

Figure 7.13 Haarlem.

Source: Topografische Dienst Kadaster, Emmen.

The urban structure of the district is fairly rigid and unalterable: the only options for change are either on a large scale, or within the existing building lines and fabric. On the other hand, the district has a spacious layout, which offers opportunities and space for change. Some of this space exists between the residential areas, but more can be created by demolition.

As in the rest of the Netherlands, average home occupancy is declining in Schalkwijk too. The sustained ageing of the population and the limited number of single-family dwellings are reinforcing the thinning-out effect in a number of districts. Most of Schalkwijk comprises flats. Many dwellings are social housing intended for lower income groups. Consequently, the main group moving to this district are first-time buyers on the housing market. There are too few houses available within Schalkwijk for residents wishing to move on, in particular to more expensive single-family homes. The disappearance of the more affluent population together with the falling average home occupancy is undermining the support for shops and other facilities. The municipality of Haarlem wishes to reverse this process in Schalkwijk, in particular by offering more residential and other opportunities for families and people with higher incomes.

7.4.2 The water issue within the urban water plan

Haarlem has always had a special relationship with water. The Spaarne is an important element in the urban structure. The Ringvaart, Leidsevaart, the fortifications, canals and *singels* are likewise prominent features in the cityscape. The municipality and the Rijnland Water Board are responsible for maintaining this water. The two organisations jointly produced the Comprehensive Haarlem Water Plan.

At about the same time, the municipality and the Water Board were engaged in developing the concept for the water network, within the framework of the Schalkwijk Implementation Plan 2000+, which is oriented to the restructuring plans for the district (Municipality of Haarlem 1999). This was used as the basis for the Schalkwijk Water Master Plan 2000+ (Municipality of Haarlem 2004b). The municipality's aim with this plan is to improve and expand the existing system of *singels* and ditches in Schalkwijk. This measure is necessary for preventing flooding, to make water cleaner, to create space for fish and plants, and to contribute to housing quality and recreational facilities. The Draft Water Master Plan was adopted by the Municipal Executive in December 2004. The Rijnland Water Board agreed to the Base Structure for the water in Schalkwijk in June 2005.

The ambition for the water system, as expressed in the Updated Schalkwijk Implementation Plan 2000+, is for a cost saving, less environmentally burdening, self-cleaning and closed water structure, i.e. a circulation system (Municipality of Haarlem 2005). This ambition can be subdivided into the following elements:

- improving the quality of the surface water;
- improving the quality of the living environment by implementing surface water (perception of nature and culture, use);
- retaining area-specific water and relieving the outlet waterway of storm water from Schalkwijk.

The execution of the master plan is divided into two phases. The first phase is concerned with the Base Structure, in which the surface water keeps its open connection with the outlet waterway. Digging will proceed for additional surface water, storm water will be disconnected for 20% of the paved area, the water quality will be improved by dredging all existing watercourses in Schalkwijk, and the northern part of the Poel polder will be redeveloped.

Parts of the water plan	Fixed	Norm surface water	Standard	Increase of surface water in the area
First draft design water system Schalkwijk	October 2003	3%	5%	9,53 ha
Master plan Water Schalkwijk 2000+	October 2004			
Integral Water Plan Haarlem	October 2004			

Figure 7.14 Overview policy documents.

Source: Municipality of Haarlem.

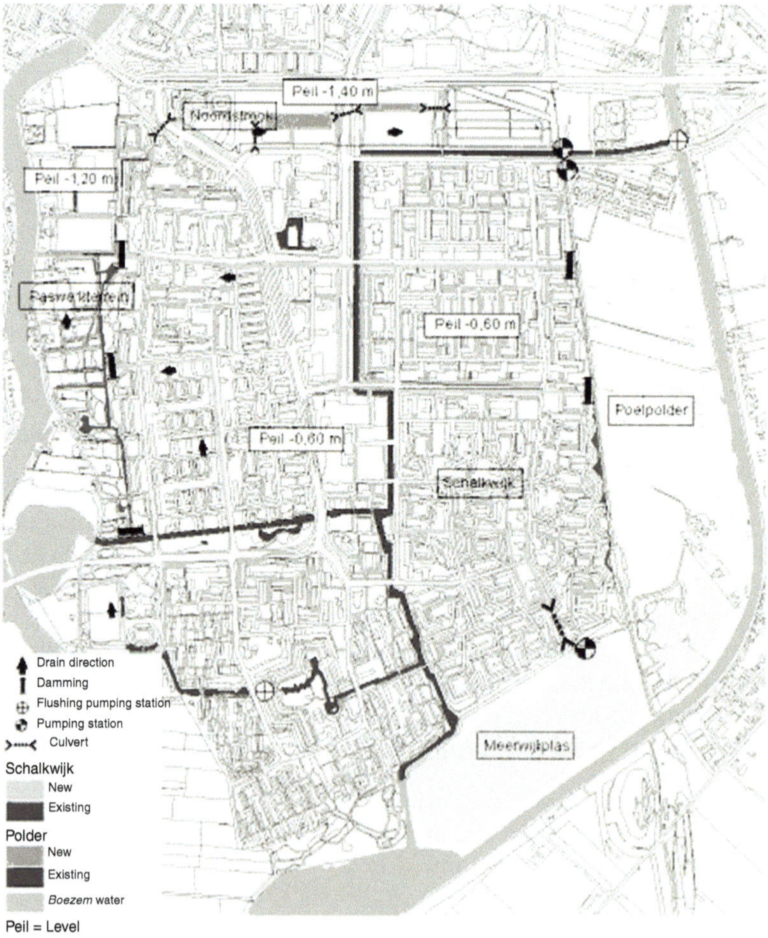

Figure 7.15 Schalkwijk Base Structure (see also color plate 24, light blue is new water; dark blue is existing open water).

Source: Municipality of Haarlem.

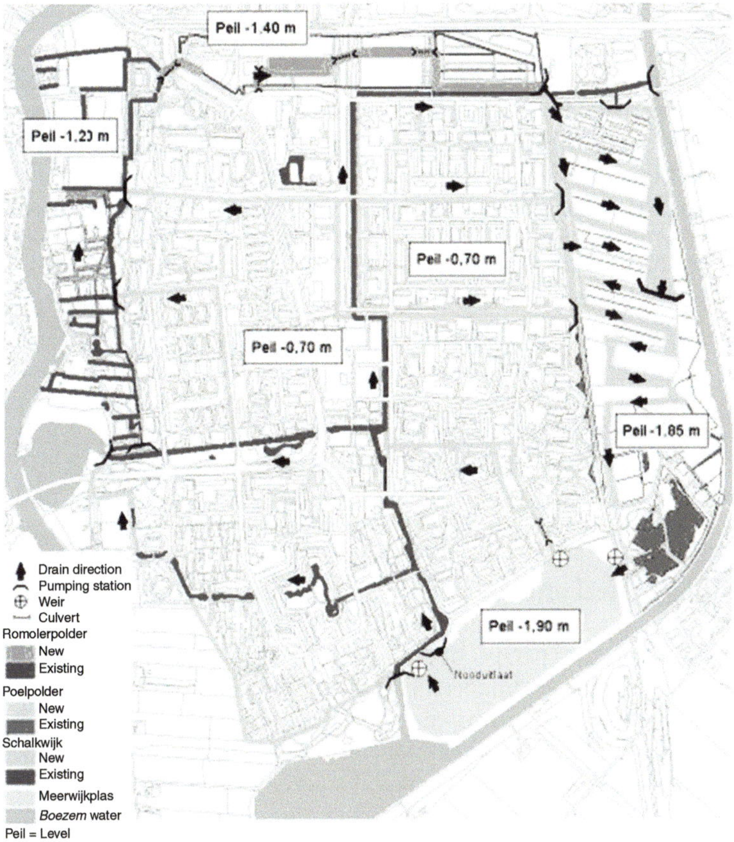

Figure 7.16 Schalkwijk Outline Design (see also color plate 25, light blue is new water; dark blue is existing open water).

Source: Municipality of Haarlem.

Once the Base Structure is complete in 2015, progress on the measures will be evaluated and the second phase, the Outline Design, will commence (Tauw 2002). The water system will then be disconnected from the outlet waterway system *boezem* and connected to natural purification fields and storage locations. If the water quality of the outlet waterway water has improved markedly in 2015, it may be acceptable from a water quality perspective not to proceed with disconnection. The waterworks in several projects are already integral components of the design. The associated implementation will take place in the coming years, and some projects have already started. Choices remain to be made regarding the water structure in other parts of Schalkwijk. Furthermore, the opportunities for disconnecting storm water from the combined sewer system have been clarified. The municipality and the Water Board will jointly identify the technical constraints for disconnection in a pilot project.

7.4.3 Solutions and process planning

An initial reference measurement performed in 1993, after lobbying from the residents for better policy and more attention to their urban district, investigated the existing state of affairs in Schalkwijk. This revealed that the greatest problem area was the population structure. The run-down condition of a number of residential complexes was a sign of the decline in the housing quality, which is determined by aspects such as the living environment, the perception of a lack of safety, too few or deteriorating facilities and diminishing social cohesion. Structural interventions were necessary in order to stop the downward spiral. The municipality of Haarlem drew up the Schalkwijk Implementation Plan 2000+, which comprises the starting points for restructuring Schalkwijk. The municipality is working to this plan together with the housing associations for the renewal of the urban district. The municipality will improve the public space, and the associations and property developers will develop, or redevelop, the property.

The water structure, one of the strategic elements of the restructuring plan, was specified in the Schalkwijk Water Master Plan 2000+. This plan was commissioned by the municipality and drawn up in close collaboration with the district water board. Representatives of neighbourhood councils, housing associations and other interested parties contributed actively to developing the Water Master Plan. A residents' festival was held as a way of creating support for the restructuring among the residents and other social organisations. The residents' festival was used as a way of drawing attention to the renewal and for capturing local knowledge of the district.

The execution of the Water Master Plan is being tackled as a project. This approach is common in renewal projects in which the water structure is developed further based on the master plan. There is liaison with interested parties and consideration is given to ambitions from other perspectives, such as traffic and housing. Further specification on a project level (spatial, financial and social) will demonstrate the actual feasibility of the water management interventions proposed in the master plan.

The implementation of the Water Master Plan envisages more than merely restoring order to water management in Schalkwijk. Water is one of the strategic elements for the restructuring and a critical success factor in achieving the desired quality boost for the living environment in Schalkwijk. The costs of implementing the Base Structure for the water are high, and adding water to an existing situation is technically complex. If the only measures taken were those demanded by the regulations, this would overlook the need for a quality boost to the living environment. The Municipal Executive considers the Water Master Plan to be a guideline for water management in Schalkwijk. Priority is then given to measures that are demanded by the regulations. When specifying the projects, the other measures (such as disconnecting the storm water from the combined sewer system) also form part of the requirements specification. The project level feasibility study then has to show whether implementation is possible. If not, the quality boost that water is supposed to give the restructuring plans is given a lower priority.

7.4.4 What does it cost and who pays?

The implementation of a properly functioning water network for Schalkwijk entails major costs. The total investment costs for the Base Structure are 45.74 million euros. The necessary resources for the first phase are drawn from multiple sources of funding, including the land development of the sub-projects, the collective operation of Schalkwijk, the sewer fund, the investment plan, the Rijnland Water Board and the urban renewal investment budget.

It is important to distinguish water measures that form part of restructuring products and those that have to be executed autonomously. Water measures that are part of restructuring can be covered at least in part from the land development of the sub-project concerned. The general operation of Schalkwijk includes an estimated water item of 2.83 million euros for the autonomous water measures. The resources from the estimated water item can be used only as counter-financing in the event of using grants from third parties. The Rijnland Water Board agreed to the construction of the Basic Structure in 2005 and made a credit of 4.69 million euros available for the 2006–2015 period. The water board is also contributing 464,000 euros to the construction of nature-friendly banks. Provincial, national and European grants are other important sources of funding for the water network. It was investigated in 2006 whether it would be possible to apply for a European Interreg grant (expected to be approximately 2 to 3 million euros).

The period occupied by the first phase is relatively long. There is a realistic probability that alternative financing facilities will yet be found. Furthermore, the water network in Schalkwijk will eventually save costs (by preventing flooding damage, reducing the quantity of soiled sludge and increasing the purification performance of the wastewater treatment plant). The Base Structure will be realised with the measures executed between 2004 and 2015.

The total costs for implementing the second phase will be approximately 22.84 million euros. The second phase is for the period between 2016 and 2025. A decision will be taken in 2015 in the light of the results achieved, the resources available and the then applicable insights as to whether to disconnect the Schalkwijk water system from the outlet waterway.

Water is valuable, but is not appreciated sufficiently in practice. The costs of this water project are clear, but the benefits and values of the water have only been outlined for Schalkwijk. Adding surface water will make the district more attractive. Furthermore, redesigning the watercourses will enhance the water experience and the water will give an extra boost to the quality of the urban living environment. This means that the water has an economic value, which must be considered when redeveloping an area. The housing stock will be altered in line with the ambitions for a more differentiated housing supply (by building homes for target groups in the middle and higher income brackets), a higher housing quality and an upgraded public space. There has also been a quest for forms of multiple use of space, in particular those combining housing, water and parking.

Not every sub-project was successful in integrating the water ambitions in full. A delay was caused because the Water Master Plan for the district was not ready in time to allow the financial aspects to be considered properly in the feasibility analyses. Integrating the water as well as possible while achieving the desired water quality and sustainability demand the creation of the watercourses set out in the master plan. The urban designer has attempted to improve the ambience of various streets by adding well-integrated surface water. This spatial quality of the water is used effectively in a number of sub-projects. The ultimate ambition was to enhance liveability by using water to provide a quality boost. The water will be unable to provide this quality boost in some other sub-projects because of the financial situation. The rise in property value is included in the allocation prices for the sites where water is being introduced. However, sufficient resources are not always available to fulfil the water ambition completely, despite the benefits being recognised.

There are various instances of land exploitation in the Schalkwijk district, and more will be opened. A risk allocation fund and a collective operation also exist, in which the foreseeable costs of district-transcending facilities have been included (e.g. water). The municipality has made strenuous efforts on various studies into the water system, disconnection options and other matters, partly in order to align with the Water Board with a view to reaching agreements on water management in Haarlem. This process has consumed much time and is one of the reasons why it now appears impossible for several restructuring plans in the execution phase to implement an important water arm. The costs of the additional surface water cannot be covered financially because they cannot be supported by the plan. There is no absolute need from a legal point of view to create more water, since it appears in the current state of affairs that the requirement for 15% surface water on an urban district level is achievable. The resources are scarce, and the ambition has duly been tempered. The option being taken in some locations is to create a green strip as opposed to surface water. The implementation costs for greenery are far lower than for water. The circulation circuit necessary for keeping the water clean is being achieved through technical means.

7.5 CONCLUSION

The many similarities between the post-war districts discussed here imply that the solutions for the water issues also resemble each other. They are also applicable to other restructuring issues.

The surplus public space and paving are being converted into new surface water in Delft, Dordrecht and Haarlem alike. Furthermore, water is used in restructuring as a means of achieving an attractive residential environment, known as 'living on the water'.

The main approach in Poptahof in Delft is of widening the existing water features at the edges of the plan area. This is a relatively inexpensive solution for integrating additional water. The presence of an existing separate water system makes it possible to offset the costs of dealing with the sewers in the restructuring project expenses. The municipality of Delft's approach in partnership with the water board and owners in the district is also worthy of imitation.

It will be hard to optimise the solution further, if only because of the municipality's active tenant participation policy. The creative solution of involving children in the development of the water plans proved to be vital. This measure avoided communication problems between the older people in this multicultural district and the municipality. Support was thus created for the water plans, with a possible positive impact on the process.

The municipality of Dordrecht attempted to strengthen the relationship between the spatial development and the Water Plan by classifying the various watercourses. They are simultaneously spatial in nature and oriented to achieving certain standards. Both the partnership on the Water Plan between the municipality of Dordrecht and the water purification board and the district water board, and the partnership with the three corporations in the district, could be copied in other cities. In particular, integrating the introduction of water into Wielwijk with the restructuring plans adds substantial value to both processes. The municipality is always at the hub in bringing together the parties involved in the restructuring and the execution of the Water Plan.

There is little probability of optimising the solution. There was attention in Poptahof too for residents' participation. This avoided the addition of surface water wrongly

Figure 7.17 New houses in Schalkwijk.

Source: Municipality of Haarlem.

Figure 7.18 New houses in Schalkwijk.

Source: Municipality of Haarlem.

being seen as the prime reason for the restructuring plans and the demolition of housing complexes.

Use was made in Haarlem-Schalkwijk of the same kind of solutions as in Delft and Dordrecht. Here too, much effort was put into residents' participation. There were residents' festivals and theme meetings on water. However, support from the residents of Schalkwijk is still only modest. On the other hand, support from the corporations is satisfactory.

The approach in Schalkwijk could therefore be improved. An important factor is that several differences of opinion have arisen between the municipality of Haarlem and the Rijnland Water Board. The process was delayed because the Master Water Plan for the district was not ready in time to allow the financial aspects to be considered properly in the feasibility analyses. There are now insufficient resources available to implement the water system in full. The implementation will now be given a lower priority in the restructuring plans. There being no absolute legal necessity to implement additional water, the water board is unwilling to contribute more. As a result, the quality ambition is being tempered, and the water will contribute less to improving the liveability of the district.

Chapter 8

International comparison

Fransje HOOIMEIJER[1] and Wout van der TOORN VRIJTHOFF[2]

[1] *Department of Urbanism, Faculty of Architecture, Delft University of Technology, Delft, The Netherlands*

[2] *Department of Real Estate and Project Management, Faculty of Architecture, Delft University of Technology, Delft, The Netherlands*

8.1 INTRODUCTION

This book deals with the Dutch situation. Is the Netherlands unique in the world in terms of the urban water management issue? Or are the Dutch solutions also applicable to other parts of the world? And might the Dutch have something to learn from solutions devised elsewhere? These are the central questions of this chapter. We have opted to focus on three countries with current projects comparable with our case studies.

The Cheong Gye Cheon River is being restored in Seoul, South Korea. Japan's technical water management approach is a good example, because it presents concepts that have been refined to adapt them to a more severe flooding issue. Examples include the Tokyo Dome, which is an ideal solution for excess rainwater in a densely populated area, and a current multifunctional project that will act as a reservoir along the Tsurumi River. The restoration of a river to a 'double deck river' is also an interesting example of the Japanese attitude. An interesting European example is the work of Atelier Dreiseitl in the Ruhr in Germany. The studio enjoys international renown for its work on the link between sustainable water systems and urban designs.

8.2 SEOUL (SOUTH KOREA)

8.2.1 Introduction

The historical identity of cities resides in the buildings and the urban structure, in which water is often a primary defining element. Historic water is acquiring a new significance as the prevailing mood is putting attractiveness above accessibility, while public support for restoring historical authenticity is growing.

Water is becoming an important structural element that defines the identity and attractiveness of historic city centres in particular. Restoration of the water structure thus largely represents an economic interest. This economic interest is translated in the scope of the local political debate into a better competitive position for the city, greater attractiveness and more visitors to the city, increased sales for local firms and rising values of existing and new property with a view over the water. Many cities are developing and implementing plans for restoring historic watercourses to their former glory, based on the above considerations and supported by ideas for improving urban water management. Many initiatives of this kind are being developed in Europe. One of the most impressive examples outside Europe is the Cheong Gye Cheon restoration project in Seoul.

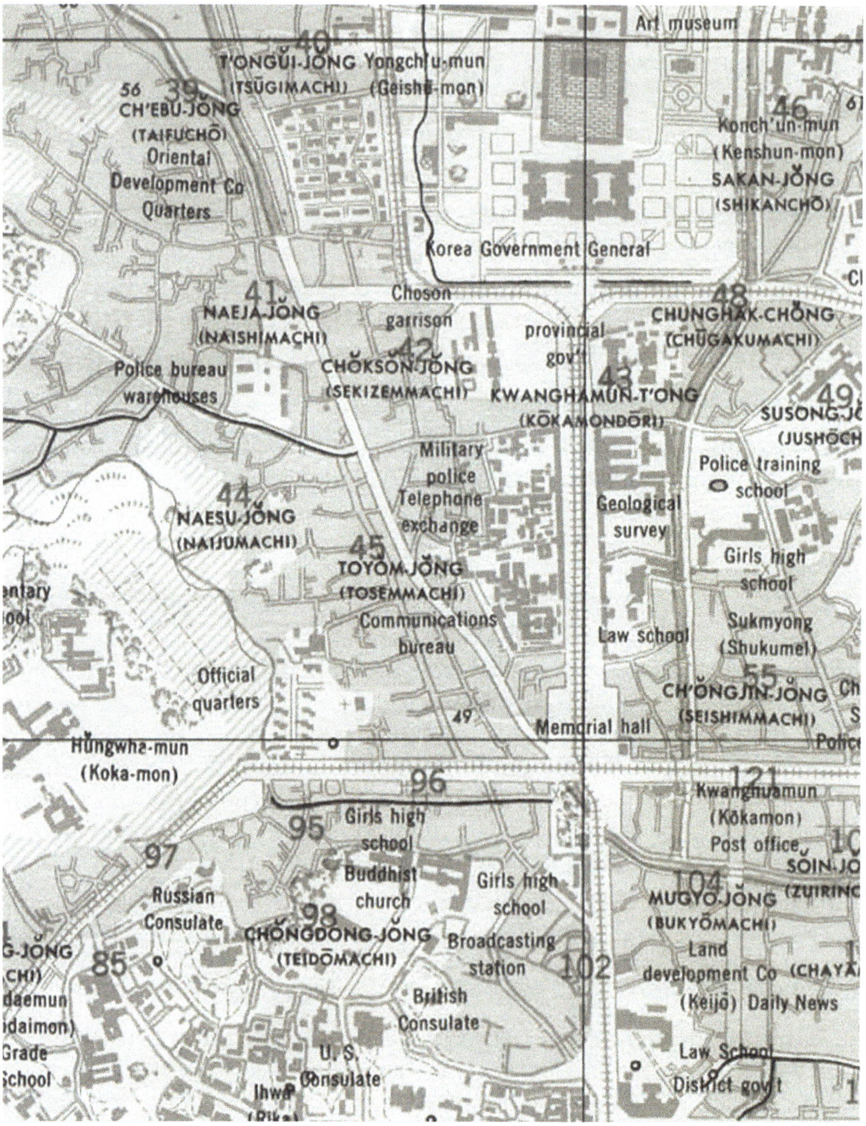

Figure 8.1 Map existing situation (see also color plate 26).

Source: Mun Jungsoo.

8.2.2 Historical development of Seoul

Seoul became the capital city of Korea in 1394 during the Choson dynasty. The city is in the valley of the Cheong Gye Cheon River, which flows from west to east through its centre. The river was regulated and quay walls were built in the urban area in the early fifteenth century, to limit the flooding that had repeatedly caused major problems.

Figures 8.2.1, 8.2.2 Seoul before and after.

Source: Mun Jungsoo.

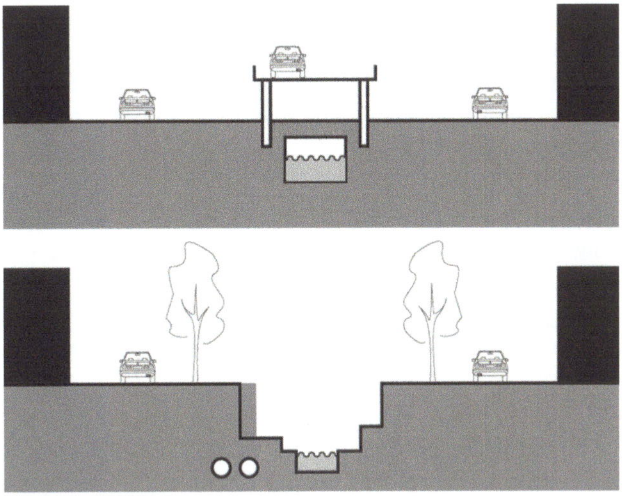

Figure 8.3 Cross-section before and after.

Source: Mun Jungsoo.

King Sejong resolved in around 1450 that the river should serve for discharging urban wastewater. However, the river was also used for washing clothes and as a playground for children. The old quay walls were unable to prevent frequent sustained flooding.

Towards the end of the nineteenth century, when Seoul had expanded into a densely populated urban centre, the flooding was causing major problems. The strongly polluted river water flooded the city and spread infectious diseases. A solution proposed for the problem was to cover the river. Only much later, in 1926, was a plan drawn up for covering the river with reinforced concrete and for developing the space for residential, retail and recreational purposes. This project was rejected by the government because of the drainage difficulties it would cause during flooding. It was the 1950s before the plan was tackled energetically, and construction completed in 1961 over a length of 2385 metres. Work then continued, so that the length of the Cheong Gye Cheon that was covered increased. The end of the 1960s saw the start of the Cheonggye Expressway construction above the course of the Cheong Gye Cheon. The expressway was 5650 metres long and 16 metres wide. The Cheonggye Expressway was a structural feat for its time and a symbol of Seoul's development.

The activities for covering the Cheong Gye Cheon were finally completed in 1977. The river that had been a defining element in urban life for almost 600 years was no more.

8.2.3 Motives for restoring Cheong Gye Cheon

The centre of Seoul, around the underground course of the Cheong Gye Cheon, developed rapidly from the 1970s onwards, supported by a fast-growing economy. The rise of an extremely varied shopping district coincided with the development of expensive apartments and offices. The hovels built by the lower socioeconomic group, which once dominated the city centre scene, quickly disappeared. A debate gathered momentum in the 1990s on restoring the Cheong Gye Cheon as a river through the centre. The following salient arguments were put forward:

- Safety. The expressway through the centre and the rapid rise in traffic intensity were causing a growing safety problem. Removing the expressway completely and restoring the river would solve that safety problem.
- Quality of the public space. Clean air, a clean river and an attractive urban space linked with the adjacent ecological park would contribute significantly to the quality of the public space in the centre.
- Culture. Seoul was built along the banks of the Cheong Gye Cheon River and the river had had vital significance for the city for 600 years. Along its banks women would do the laundry, children would play, and merchants would conduct their business. The restoration of the river would link the city with its past again. The river and the nine ancient bridges would contribute significantly to strengthening the cultural identity of the city, which now has 20 million residents.
- Economic boost. The marked improvement in the quality of the public space is expected to strengthen and intensify economic activity in the central area.

The plans were drawn up and executed with support from the mayor of Seoul elected in 2002, Lee Myung Bak. Before Mr Lee was mayor, he was head of Hyundai Engineering and Construction, which was one of the parties involved in the construction of the expressway over the Cheong Gye River. On a visit to the underground river, Mr Lee came to understand that the river had to be restored to its former glory. *Sunlight filtered through cracks in the tunnel roof. In the light of one, Lee recalls noticing a spot on the ground where a melon seed had taken root and a tiny flower had blossomed. Lee says that he 'was deeply impressed and it brought me to tears'.*

8.2.4 Main features of the plan

The demolition of the constructions above the original riverbed started in mid-2003 and was completed at the end of 2004. Much of the 680,000 tons of demolition material was recycled.

Plans to supply water from the original springs have been postponed for the long term because they proved infeasible in the light of traffic problems and the expense involved. Furthermore, the water level in the Cheong Gye Cheon normally varies considerably, so that the bed is dry for much of the year. Now 120,000 tons of water a day has to be brought in to the start of the reopened riverbed in the urban area to guarantee flowing water. The water comes mainly from the river Han into which the Cheong Gye Cheon flows.

The quay walls built along the river are high enough to withstand all but a water level that is likely to occur only once in two hundred years. Large sections of the riverbanks are accessible to pedestrians. Five bridges are reserved for pedestrians and seventeen for motorised traffic.

The discharge of wastewater is through pipes beneath the riverside roads. This sewer system was dimensioned for the discharge of 1.95 million tons of wastewater a day.

The project costs are estimated at more than 700 million euros, to which approximately 9 billion euros should be added for the redevelopment of the 792 thousand square metres of commercial and residential development over the coming five years.

8.2.5 Learning from South Korea

The plans for Seoul, which are now complete, were extensive and drastic. Legal procedures would probably considerably extend the proceedings if a similar initiative were to be taken in a major European city. The improvement of urban water management was only a minor argument in justifying the plans.

The project was presented in the market as the restoration of an important historic watercourse. In Seoul too, there was some criticism of the concessions made on authenticity.

Lee Joo Yeon, the editor of the local architecture journal *Space* does not view the project as a restoration. He points out that it is supplied with water from a nearby river and water treatment plants, not from the original springs. '*This is just an artificial waterway being constructed in the city; it is not a historic restoration.*'

Ignoring the scale, the Cheong Gye Cheon project has many parallels with the projects in Breda and Utrecht. The main similarities are:

- the restoration of historic watercourses;
- the importance of the watercourses for the identity of the cities concerned.

A historic watercourse was opted for, taking current technical, hygienic and aesthetic requirements into consideration, which implies that the result is not an exact replica of what existed 100 years ago

The contribution made by restoring historic watercourses to improving urban water management is small, and to improving the quality of the public space large

The local government took responsibility for the investments in the public space and expected to stimulate private property investment in the immediate surroundings as a result.

8.3 TOKYO (JAPAN)

8.3.1 Introduction

Japan consists of mountains and polder land. The rough and densely forested mountains are almost uninhabited and the polder land is organised almost exclusively as rice paddy or city. Japan has no urban design, and no large-scale plans for streets, squares and houses with gardens. Instead, the cities expand stepwise by developing on rice paddies. The fields are raised, drains installed and a street laid, and houses built as tightly together as possible on both sides of the street. The cities fan out in this way and merge into paddies with the occasional couple of houses or a fully developed field. Villages as such do not exist.

The reason for the absence of large-scale planning is the power of the private landowners. The possession of a paddy is vitally important for a family. Family possession is sacrosanct and is how people make a living – the price of rice in Japan is kept artificially high – and where they live. It is therefore almost impossible to develop plans of any size, and the stepwise expansion means that paddies persist within the city because families refuse to build on or sell them.

A major advantage of this urban expansion without an urban plan, compared with the Dutch situation, is that it retains the national water systems, which are a conspicuous part of the city. There is therefore none of the meddling with the water system that is partly responsible for our current water problems. The rain in Japan is heavier, concentrated in

Figure 8.4 Aerial view Japanese polder landscapes (see also color plate 27).

Source: Fransje Hooimeijer.

shorter periods and of twice the quantity as in the Netherlands, but they have no flooding problem precisely because they have retained the national water system.

The water system has not been retained everywhere. Japan's largest water city, Tokyo, was once mentioned in the same breath as Venice, Amsterdam and St Petersburg. It is still possible to visit Tokyo's origins, the castle town of Edo, or 'the high city', situated on a ridge, and look out over the adjacent delta where 'the ordinary folk' lived in the 'low city', reclaimed from the water. Like Amsterdam, canals were dug in the low city to drain the land and there were also harbours and warehouses. This was the site of most of the business activity and the city's entertainment. Floating theatres and ceremonies on the water were important cultural aspects that made life in the low city extremely popular.

This city was devastated by an earthquake in 1923. The municipal administration then had a unique opportunity to implement a new street plan, suitable for the car. The system of stepwise expansion simply continued and has yielded a fascinating cityscape full of differences in scale and functional combinations. This mosaic is normal for the Japanese, and you never hear them grumble about the nuisance of highways through the city. The barrier to large-scale plans did become a problem when Tokyo was selected as the venue for the 1964 Olympic Games. The city wanted to put on a fine show on the world stage with this international event and introduced a modern highway system. The power of private land ownership formed a barrier, and the only way around it was the water. You regularly encounter elevated highways standing on

Figure 8.5 Nihonbashi Bridge (see also color plate 28).

Source: Fransje Hooimeijer.

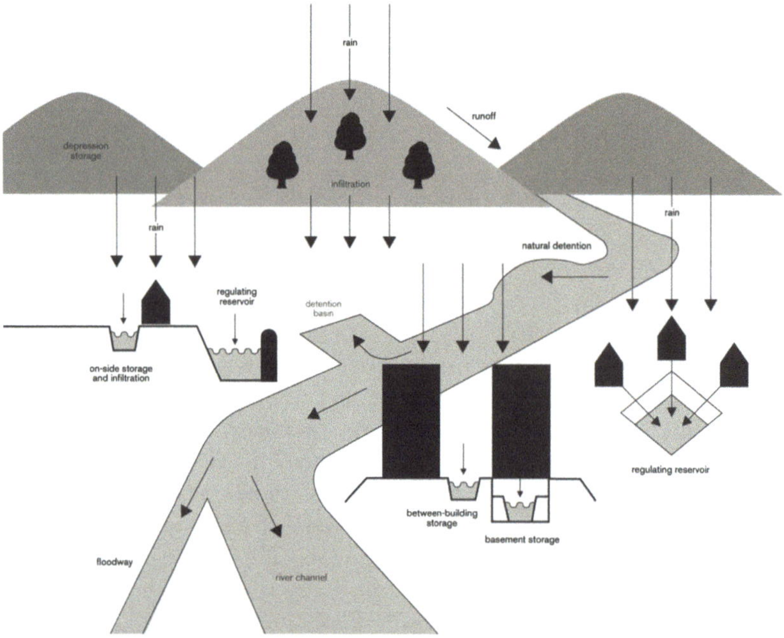

Figure 8.6 Various Japanese solutions.

Source: Arakawa-Karyu River Office 2004.

pillars in rivers and canals in the city, or catch a glimpse under an old bridge of traffic flashing by on a highway on the bed of what was once a watercourse.

The water city is concealed behind enormous concrete pillars and flyovers, but has not been lost. Tokyo's most important historic bridge, the Nihonbashi, with its ancient features, simply plays its part in the ultra congested traffic, and has been covered by an enormous expressway. Just as historic watercourses are being dug out again in the Netherlands, the Japanese are also interested in Tokyo's historic identity, and are planning to make the bridge 'expressway free'.

Like the Netherlands, an ecological revolution took place in Japan in the 1970s after the oil crisis, with greater regard for the quality of water. The awareness and the responsibility of the general public for the environment is impressive. Both land and water are spotlessly clean, and they have a variety of projects and methods to keep them so. An example is the toilet that when flushed causes water to flow into the washbasin through a tap, from where it goes into the cistern for reuse in the next flushing cycle.

Japan has had a law since the 1980s related to flooding and water quality that obliges large buildings, such as the Tokyo Dome, to capture storm water, process it in an internal water treatment system and reuse it for flushing toilets, for example. Dutch ministers and study groups flock to Japan to scrutinise examples of this kind and see what might be used here.

The Japanese have largely absorbed the hydraulic engineering knowledge of the Dutch. Dutch engineers have had much impact in Japan, in particular in river construction,

Figure 8.7 Tokyo Dome.

Source: Fransje Hooimeijer.

and notably Johannes de Rijke is a national hero in Japan. Lakes were drained in Japan after World War II, following the Dutch model. One drained lake north of Akita reclaimed from the sea could be the twin of the Haarlemmermeer. However, as in other areas, the Japanese have surpassed us. They are building islands in Tokyo Bay and Osaka, which are many times larger than the Maasvlaktes or IJburg, and which have enormous shopping malls and residential and office buildings.

8.3.2 Tokyo Dome

There were three external factors in the early 1970s that changed the political climate and the planning of environmental regulations abruptly. The first was the opening of diplomatic relations with China, the second was abandoning the dollar exchange rate peg and the third was the oil crisis of 1973. Japan's lack of energy resources (in the form of oil and gas) makes it heavily dependent on import, which was painfully obvious in the 1970s. They started to focus on more efficient use of fuels and recycling.

Regarding water facilities, they introduced the rule that a building larger than 10,000 square metres must have internal water recycling measures, in accordance with a '*stick and carrot*' approach. While a developer is obliged to install facilities, the carrot takes the form of favourable interest for loans from the Development Bank of Japan.

The Tokyo Dome, the home base of the Tokyo Giants baseball team, is a good example. Storm water running off the roof of this building (16,000 square metres) is treated and used as 'grey water' in the kitchen, showers and for flushing toilets. A 3,000 cubic metre reservoir was installed under the stands; 1,000 cubic metres were for use as flushing water, 1,000 cubic metres for fire extinguishing and 1,000 cubic metres reserved for receiving storm water. This storm water facility has almost eliminated the consumption of water for flushing toilets and washing, which until last year amounted to hundreds of cubic metres. The building also has an internal water treatment plant, which 'purifies' the used rainwater once more for reuse as grey water before finally being discharged into the sewer. Another important reason for the rule was to stop groundwater extraction for drinking water, which was causing subsidence.

8.3.3 The Tsurumi River multifunctional project

The Tokyo Metropolitan Government and the city of Yokohama have produced a master plan for an overflow location with multifunctional facilities for dealing with the peak discharge of the Tsurumi River. The land in the extremely densely urbanised area was adapted for the water issue such that it could be used for water storage in periods of high incoming water volumes. The rest of the time it is a sports centre.

The area is 84 hectares with a capacity of almost 4 million cubic metres of water. The overspill dike for peak flow is 777 metres long with a capacity of 200 square metres a second. When the plan is complete, the overflow capacity will be 800 square metres a second, which includes the capacity of the river forelands and the other upstream overflow areas.

The plan includes the Yokohama International Stadium, which can accommodate 72,000 people (the number of spectators at the 2002 World Cup final), a medical

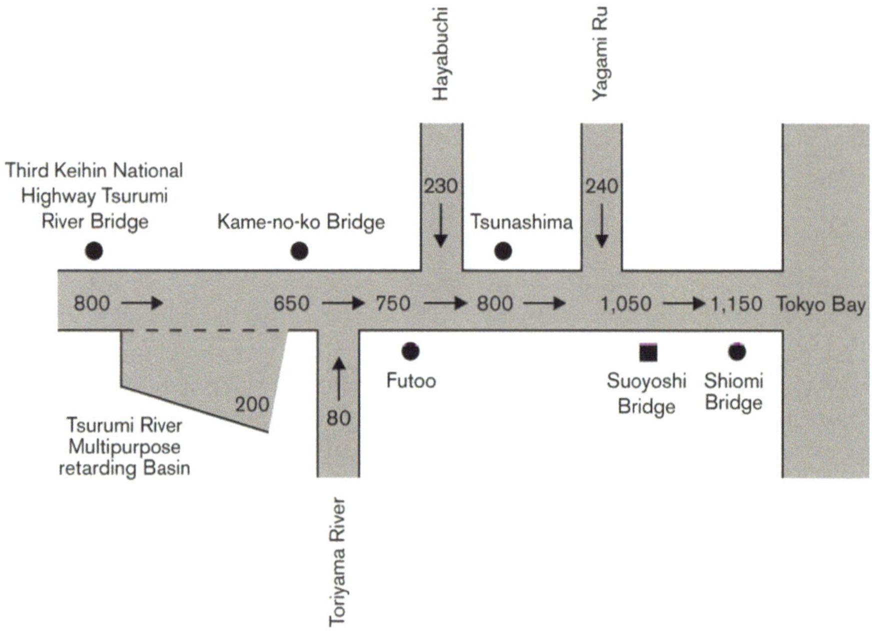

Figure 8.8 Flow Rate Distribution.

Source: Arakawa-Karyu River Office 2004.

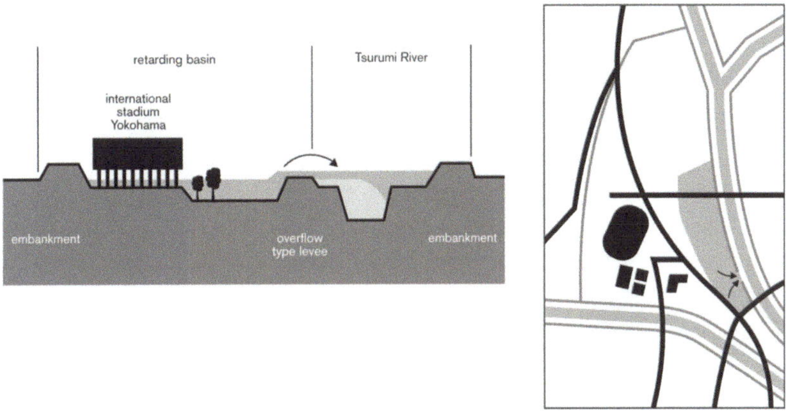

Figure 8.9 The river is allowed to overflow at the overflow location in the event of flooding.

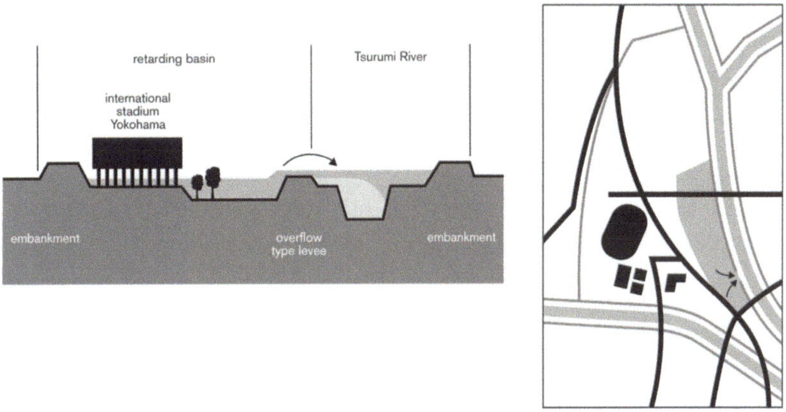

Figure 8.10 The water is stored temporarily at the overflow location.

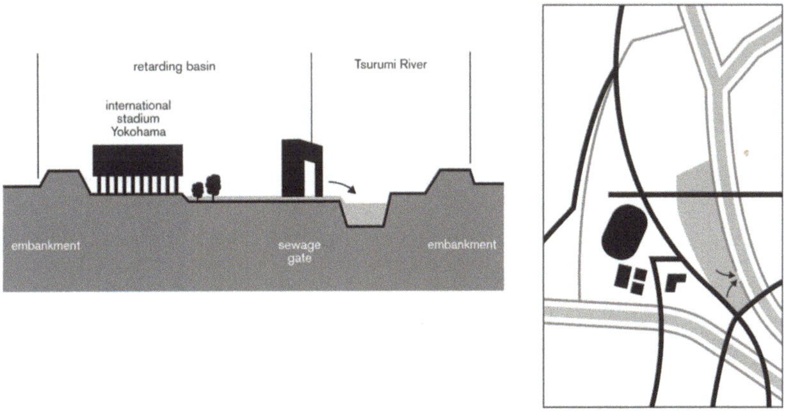

Figure 8.11 When the river water level has returned to normal, the water can be discharged back into the river.

Source: Arakawa-Karyu River Office 2004.

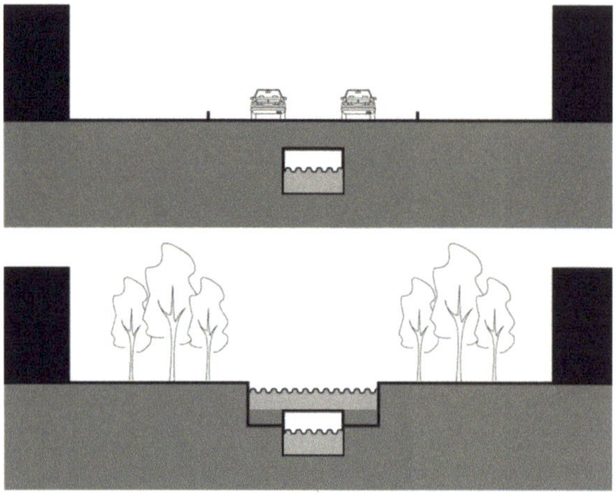

Figure 8.12 Double deck river.

Source: Fransje Hooimeijer.

centre, a rehabilitation centre and a sports and cultural centre for disabled people. The Yokohama International Stadium and the Yokohama Comprehensive Care Continuum have been built with a raised floor to allow water to flow underneath.

8.3.4 'Double-deck river'

Another factor after the ecological shift in the 1970s was that water started to be seen as part of the public greenery, which is scarce in Japan. For instance, the Kitazawa River has been brought back above ground at the request of the residents of a Tokyo neighbourhood near Ikenoue train station. The Kitazawa River flowed underground through two large conduits and its main function was as a sewer. The district residents wanted the water to be visible again in the neighbourhood. The municipal government therefore created an elegant stream on top of the conduit, in consultation with the residents. The stream is fed from the outflow of a purification plant 17 kilometres (!) away. An important aspect is that each local neighbourhood of 100 to 200 residents has produced its own design for their part of the stream, which has created extraordinary diversity. This is a good example of tenant participation, and the successful demonstration is prompting many more neighbourhoods in Tokyo to copy this facility.

8.3.5 *Superlevee*

A start has been made along the Ara River in Tokyo with the construction of *superlevees*, which are extra high, broad dikes with an inside slope of 1:30, on top of and within which buildings, roads and other facilities are constructed. The toe of the dike

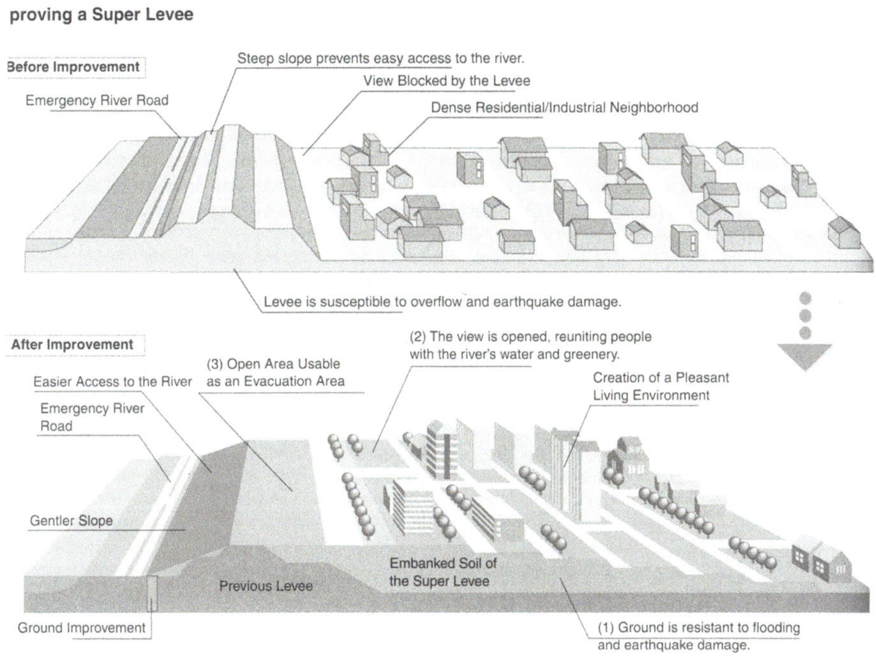

Figure 8.13 Superlevee (see also color plate 29).

Source: Arakawa-Karyu River Office 2004.

is reinforced underground with steel sheet piling and internally with a concrete slab. The dike body consists of soil rather than sand. The Ministry of Land, Infrastructure and Transport (MLIT) is transforming sections of dike – some of which are only 200 to 300 metres long – into *superlevees*. They are working closely with the Tokyo Metropolitan Government Spatial Planning Department. The Tokyo Metropolitan Government is negotiating with the landowners and homeowners behind the dike and is either buying them out or arranging temporary accommodation so that they can return after the *superlevee* construction. Rather than uncoordinated building, there will be a park and high-rise building.

It is hoped to gradually create *superlevees* along the entire river from the estuary to 30 kilometres upstream. People do not mind that this process may take many decades with the current approach; it is simply the way it is. An important advantage of the *superlevee* is that building is possible on the dike and the land can therefore be used. The high Tokyo land prices make this an extremely significant point. It also avoids people being confronted with a view of a high dike.

Efforts are also being made to use water in the river basin. Seventeen river basins have been identified for integrated flood management, which is common in particular in the Tsurumi River near Yokohama. Tokyo also has underground floodways to enable water to be pumped out of the city. These pipes are more than big enough to accommodate an underground railway.

8.3.6 Learning from Japan

It is interesting to see how they are dealing with comparable water problems in Japan. Japan is demonstrating what is possible if you simply make a start and are willing to try out newly available techniques on a small scale, to improve them, and to apply them on a larger scale. They also have considerable faith in their ability to solve complex problems with large future uncertainties.

A point particular worthy of adoption by the Dutch is the boldness to start a project without knowing for certain whether all of it will ultimately prove feasible. Boldness of this kind is what our ancestors possessed when they created this country from the water, but we have lost somewhere along the way. Constructing a *superlevee* 200 metres long without knowing whether the other 27 kilometres will be forthcoming is a clear example of this boldness. What would be said immediately in the Netherlands is that the 200 metres is pointless without the other 27 kilometres, whereas in Japan they are just convinced that it will work. We could apply a far more stepwise approach to our water problems, by making incremental interventions that will ultimately produce a safe and dry Netherlands.

8.4 THE RUHR (GERMANY)

8.4.1 Introduction to Atelier Dreiseitl

Atelier Dreiseitl was incorporated in 1980 by Herbert Dreiseitl with the objective of developing sustainable projects with an aesthetic and social value. The studio is a multidisciplinary practice specialised in applied art, urban hydrology, ecological technology and landscape architecture. The projects have a particular focus on water, which is one of the most vital and threatened natural resources on earth. The aim is to set a process of awareness in motion and to demonstrate that sustainability and good design solutions are mutually compatible and interdependent.

The scope of the work extends from integrated sustainable storm water designs for urban parks, to water features, swimming baths, water playgrounds and residential building projects.

8.4.2 Restoration of the Emscher

The Emscher is a relatively small tributary of the Rhine that runs through the Ruhr in North Rhine-Westphalia to the Netherlands. It is 84 kilometres long, and the average discharge into the Lower Rhine is 16 cubic metres a second. The source of the river is in Holzwickede, to the east of Dortmund, and it flows past the cities of Dortmund, Castrop-Rauxel, Herne, Gelsenkirchen, Essen, Bottrop, Oberhausen and Dinslaken. Before industrialisation, the Emscher and its tributaries were a meandering water system on a gentle slope. The area was entirely paved in the course of industrialisation, and the water had to be pumped away through an artificial system, which completely destroyed the original landscape.

As the geographical centre of an industrial zone with 5 million residents, the Emscher is biologically dead. By the end of the nineteenth century, it had degenerated into an open sewer. A doubling of the built-up area in the Emscher zone between 1954 and 1981 has correspondingly doubled the nuisance. The probability of flooding

Figure 8.14 Masterplan Emscher Park 2010 (see also color plate 30).

Source: http://www.elp2010.de.

increases during heavy rainfall, because the water cannot be stored in the hard surroundings and seeks a way out through the Emscher. Sedimentation and the mining industry in the surroundings have also caused the natural course of the river to be altered twice in recent decades. An enormous water treatment plant at the confluence with the Rhine now guarantees a certain water quality at that point.

The Emscher Park International Building Exhibition (IBA) had a structural programme oriented to boosting the improvement of the Emscher zone. The most important objective was the urban development and ecological renewal of this highly polluted former industrial and mining area. The programme started in 1989 and was completed successfully in 1999.

The ecological renewal of the Emscher and its tributaries was combined with the construction of new and modern sewer and drainage systems. Another significant intervention was the construction of a green corridor to link all seventeen cities in the zone through the existing watercourses and the Emscher Landscape Park. The foul, black stinking sewer that once ran straight through the Emscher Landscape Park was dealt with. The restoration of the Lanferbach started with the separation of water running off the adjacent slag heap, Monte Schlacko, and the sewer channel. A large pipeline has been installed on the river site for the sewer and storm water. A 51 kilometre long concrete pipe between Dortmund and Dinslaken, between 10 and 40 metres below ground, is scheduled to guarantee the drainage in the Ruhr area in 2019. The water draining from the renovated and expanded Schüngelberg residential area is captured in a designed network of infiltration fields, where the water is cleaned before being returning to the Lanferbach.

8.4.3 Project organisation

The IBA Emscher Park Planning Company Ltd was incorporated to supervise and finance the 120 IBA projects. The 3 billion euros funding of the total project has been arranged through local, regional and national public authorities, the European Union, the commercial sector, NGOs and PPP constructions (Shaw 2002).

Figure 8.15 Plan River Volume (see also color plate 31).

Source: Atelier Dreiseitl.

Figure 8.16 Town Hall.

Source: Atelier Dreiseitl.

The plans were produced with input from national and international professional experts and implemented under the responsibility of the steering committee with representatives of the local public authorities. The steering committee also included contractors and developers. Public participation was considered crucial and of constant importance. The public was involved in the project through festivities and exhibitions in order to encourage this participation.

Figure 8.17 Town Hall's coloured water wall.

Source: Atelier Dreiseitl.

8.4.4 The restoration of the Volume creek

Atelier Dreiseitl restored the Volume creek in Hagen, also in the Ruhr. This river degenerated into an open sewer for the same reasons as the Emscher. It is concealed behind an industrial complex and polluted with industrial effluent. This might actually be why the city of Hagen decided to build the new town hall on this site, in that it presented an opportunity to transform the river into an attractive public space in the city. For the designers it was an initial symbolic step in restoring the region's hydrological system.

Herbert Dreiseitl saw it as a challenge to create a visual link between the new town hall and the river. He did so by effectively bringing the water into the building by introducing partly transparent paving. He also made a coloured water wall onto which the water trickles and flows down. Moreover, the water captured on the roof meanders down over cascade stones to the river.

8.4.5 Learning from Germany

Although nothing to do with regulations for water, the principle behind the German International Building Exhibition (IBA) has proved to be extremely effective in regenerating run-down areas. Focusing on an issue and then devoting all resources to finding a solution in a consortium has been shown to be fruitful. It is more than a financial catalyst, and by opening the area to the public as an exhibition involves people in spatial problems and gives them insight into the solutions.

Atelier Dreiseitl's work is an example for projects involving a combination of a spatial issue and a water issue. The water issue is resolved integrally with other spatial issues and thus adds value. In other words, the point is not stopping with the restoration of a river's ecology, but also making it a functional part of a residential area, as both projects demonstrate.

Chapter 9

Conclusions

There are three aspects to the problems caused by climate change that the Netherlands is facing. As is now generally accepted, the earth is heating up, with human behaviour partly to blame, and causing sea levels to rise. In putting a figure on the magnitude, a maximum of approximately 85 cm in 100 years is being assumed. Guaranteeing the survival of the Netherlands will involve strengthening and raising the existing coastal protection. The measures involved will have substantial financial consequences, although they have strong public support and are not politically controversial.

Secondly, problems are being caused by increasing precipitation, which furthermore is tending to be concentrated in brief periods, beyond the capacity of the major rivers for which the Netherlands is the delta. The solution is found in raising the river dikes, but in designating overflow areas along the major rivers. These areas can be allowed to flood when the rivers rise above the river dike design level. The need for areas of this kind fortuitously coincides with an increasing surplus of farming land in the Netherlands.

Thirdly, incidental heavy rainfall, in particular in urban areas, is causing acute problems. The urban water system was not dimensioned for the peak loads that have occurred in recent years, and will continue to occur in the future. The measures that can be taken to tackle the problems in the urban area are usually, but not always, combined with spatial interventions.

The most convenient, and probably the least expensive, solution is to temporarily lower the surface water level by preventive pumping. The surface water level is then lowered by between 10 and 20 cm a few days prior to heavy rainfall. This measure creates storage capacity without requiring spatial interventions. However, even if the forecasts are reliable, the measure still has shortcomings. The rainwater does not then remain on the street, but runs into the combined sewer, which consequently overflows. Furthermore, it will involve pumping away much clean water that will be needed later, especially in the summer. The only supplementary supply will be river water that is polluted or from outside the area.

Increasing the storage capacity is another possible solution. One way of doing so is to create more surface water in the form of ditches, canals and *singels* and by restoring urban watercourses that were previously filled in. It is also possible to create underground reservoirs, or urban buffer zones, by analogy with the buffer areas along the major rivers. One possible form of the last measure would be the regulated flooding of urban squares, the ground level of which is duly lowered for the purpose.

Finally, modifying the underground network of pipes can increase the discharge capacity.

The case studies presented in this book show various examples of the relationship between the spatial interventions, the technical measures, the necessity related to the water issue and the motivation for the project. The Poptahof in Delft actually has too much open space in the urban fabric and the water issue is being used as an opportunity to improve the quality of the public greenery concerned. Financing the water issue can also be incorporated in the general costs, in an extraordinary construction in which red really pays for blue!

A public space quality improvement approach of this kind was also appropriate for Wielwijk in Dordrecht. The municipality is paying for the water issue, and the project developer concerned is contributing to water structures that will raise the market price of adjacent homes. The project developer will be able to put these homes onto the market at a price 10% higher than those not on the water. Although the district water board initially agreed to contribute, they withdrew at the last moment. Is this a sign of unclear regulations? The water issue here consists of water collection during peak precipitation. However, the water requirement in dry periods also demands seasonal storage, which has not been addressed here. It is nonetheless worth asking whether there might have been opportunities to do so in combination with the park of the Dordtse Hout.

The funding of the restructuring of Schalkwijk in Haarlem and of the water issue (adding much more surface water) were covered from separate sources. Restoring the water structure, as opposed to a structure that runs alternately above and below ground, is sometimes used to emphasise the restructuring. This is because, as in the above examples, it solves the problem of too much public space. It is a good example of the application of the toolbox of guiding models discussed in Chapter 4, and an integration of qualitative and quantitative water issues. The circulation model makes Schalkwijk a separate system, which is to say separate from the outlet waterway. This is not an unnecessary measure if the quality of *boezem* water improves.

An example of a solution that does not directly use urban space but falls under the category of multifunctional use of space, is the multi-storey car park-cum-water storage facility under Museumpark in Rotterdam. The municipality is paying two-thirds and the water board one-third of the 6 million euro bill for the multi-storey car park and 10,000 square metres of water storage. The simplest and least expensive way of storing additional water is in surface water, although the question may arise as to why the municipality did not take this option in Museumpark, which is a park with much space. Not only is the selected solution more expensive (600 as opposed to 350 euros per cubic metre), but the storm water still flows into the sewer and to the treatment plant, which are not the desirable water solutions. The reason for opting for the multi-storey car park was that Museumpark already fulfils a complete space claim, with the museums and a classical garden that the park contains. It would be impossible to add 10,000 m3 of water. Furthermore, the multi-storey car park is one of the first multi-functional retention reservoirs in the Netherlands!

There is another quality improvement that simultaneously benefits the spatial quality in the Vogelwijk in The Hague, albeit that no spatial intervention was involved. For example, no attention was paid to the course of the stream water and whether it might be newly laid out as a pleasant walking route. The lack of a presentation of the project makes it invisible to the public and results in meagre public support. Water quality is specifically in the water board's portfolio, but the municipality is also paying some of the costs in this case.

Utrecht is an excellent example of a hydraulic engineering intervention that also restores the urban design logic. Restoring the *singel* is making the urban fabric comprehensible again, in particular because the original buildings are largely intact. The intervention has a symbolic value and no influence in the short term on improving the water management (which will, however, happen in the long term). This is why the municipality alone is responsible for the costs. The project demonstrates that various parties with separate interests do not always arrive at a clear solution, and in this case the distribution of costs is not logical and other interests that remain vague must be playing a role. Suppose that the reopened Catharijnesingel is also a cost-efficient measure for water management. It would then be logical for the water boards to contribute. The fact that they are not would suggest that they believe the water issue could be solved at less expense. However, if other reasons also exist in favour of the *singel* plan, it would be logical for the water boards to pay a proportional amount. After all, 'their' problem is being solved! On the other hand, water is deemed an improvement to the living climate in the city centre. This is an important point for the municipality, but the other partners also benefit, while it is unclear whether they are helping pay for reopening the *singel*.

The absence of many of the original buildings and the fact that the original function is not being restored means that Breda's Old Harbour restoration is not contributing to improving the logic of the urban fabric. This project is a form of urban revitalisation, refurbishing a poor district and an economic boost to the city new style. This is also why the project is being paid for by the national government and the municipality. The project provides no solution for an urgent water issue and has no ecological starting points.

Finally, isolating the water system of the east end of Delft city centre protects the city centre from flooding. The solution is a technical rather than a spatial one, which can uphold Delft's extensive water structure. A spatial solution would mean that all the quays would have to be raised, which is unthinkable in a slowly subsiding city centre. The costs are being borne jointly by the municipality and the water board.

Besides quantity, there is also a struggle to combine the water issue with a quality improvement in the urban public space. The case studies reveal that the district water boards often fall back to what is legally necessary and opt for the most cost-efficient and technical solutions. While often a more expensive solution than surface water (see Museumpark), choices for redesigning public space by expanding the surface water are probably also more complex organisationally and in terms of funding, which is probably another reason not to start. If this solution is chosen, the district water boards tend to withdraw, while there is no guarantee of success in persuading the developers and corporation to participate in an equalisation strategy. Nonetheless, there is cause for optimism in the case studies.

At a high level, these Dutch examples share characteristics with the case studies from abroad. Although the water issue in South Korea is to oppose flooding, which is an important starting point, this aim can be achieved only with a major spatial intervention, which also benefits the urban quality in general. A water issue is not always necessary in Japan for water to be brought back into the city; it is seen as an urban quality, like a park or a playground. On the other hand, the Japanese are better at integrated solutions. The Netherlands only has a multi-storey car park, while Japan has had projects like the Tokyo Dome since the 1970s. Japan is also more progressive in opting to start small and then move on to tackle the entire project in steps. Projects in the Netherlands are required to serve a larger use from the outset. The same is evident in the examples from Germany, in which the ecology of the city, in relation to the quantity of water and the quality of the public space, is comparable with the post-war Dutch projects.

It is clear that spatial interventions cost time and money. Time, because the total urban water management system cannot be renovated at once. It has to happen in steps, and preferably in combination with an approach to the property in need of improvement or replacement. If this combination can be achieved, the costs for the alterations for water storage and the underground network of pipes can be incorporated into the total plan costs. Otherwise, the costs will have to be covered from local and regional taxes. The objective formulated in accordance with policy is that the cities in the Netherlands must have their water system in order by 2015. This does not yet mean that the total urban water system is to be completely renewed. The phasing of the renewal will depend on the opportunities for covering the costs from area-related land developments and local and regional taxes. Spreading the implementation over a long period makes the scale of the upgrade clearer and more affordable at the same time. We should realise, however, that the current urban water management system is the result of centuries of development, involving a combination of experience and technical ingenuity. The Netherlands has an urban water management system the capacity of which must be increased somewhat, which is an entirely different matter from developing a completely new system. The Dutch have the task of modifying their inheritance, which is both technically feasible and affordable. In other words: we are going to do it.

Bibliography

REFERENCES

Arakawa-Karyu River Office (MLIT). 2004. *The Arakawa: river of the metropolis. A comprehensive guide to the lower Arakawa.* Tokyo.

Beck, U. 1993. *Risk society: Towards a new modernity.* London (Sage).

Bervaes, J.C.A.M. and Vreke, J., 2004. *De invloed van groen en water op de transactieprijzen van woningen (The influence of greenery and water on the transaction prices of homes).* Alterra report 959. Wageningen.

Bos, A. 1946. *De stad der toekomst, de toekomst der stad (The city of the future, the future of the city).* Amsterdam (Voorhoeve).

Breebaart and Huffstadt, with KCAP. 2006. *Studie Bouwprogramma Rotterdam-Zuid (Building programme study for South Rotterdam)*, Rotterdam.

BRO.2000. *Water in de Westflank van de Bredase binnenstad: economische effecten en aanbevelingen (Water in the Westflank district of Breda city centre: economic impacts and recommendations).* Amsterdam (BRO).

Bureau Parkstad. 2004. *Richting Parkstad 2015 Ontwikkeling voor de vernieuwing van Amsterdam Nieuw west (Towards Parkstad 2015. Development for the renewal of Amsterdam's New West district).* Amsterdam (Bureau Parkstad).

Burke, G.L. 1956. *The making of Dutch towns. A study in urban development from the tenth to the seventeenth centuries.* London (Cleaver-Hume Press).

Claringbould, M. and Hinterleitner, J, 2005. *Perron* H_2O. Wageningen (Alterra).

De Greef, P. and Urban Design and Public Housing Department. 2005, *Rotterdam Waterstad 2035 (Rotterdam Water City 2035).* Rotterdam (Urban Design and Public Housing Department).

Delftwonen. 2003. *Werelds wonen in de Poptahof (Worldly living in Poptahof).* Delft (Delflandwonen).

Doorn, M. van and Mennema, J. 1992. *De Haagse Beek. Een (natuur)historische verkenning (The Haagse Beek. A natural history survey).* The Hague.

Eesteren, van C. 1934. *Algemeen Uitbreidingplan (General Expansion Plan).* Amsterdam (Municipality of Amsterdam).

Hereijgers, A. and Velzen, A. van, 2001. *De naoorlogse stad. Een hedendaagse ontwerpopgave (The post-war city. A contemporary design issue)* Rotterdam. 2001

Hooimeijer, F.L., V.J. Meyer and A. Nienhuis. 2005. *Atlas van de Nederlandse waterstad (Atlas of the Dutch water city).* Amsterdam (SUN).

IPCC. 2001. *Climate Change 2001: The Scientific Basis, Intergovernmental Panel on Climate Change.* Cambridge.

Kaap3, De Beek and Van Gilst. 1999. *Toekomstvisie Rotterdam Zuid 2030* (*Vision for the future of Rotterdam South 2030*). Rotterdam (Urban Design and Public Housing Department, Rotterdam City Development Corporation).

Karsten, L., A. Reijndorp and J. van der Zwaard. 2006. *Stadsmensen. Levenswijze en woonambities van stedelijke middengroepen* (*City people. The way of life and housing ambitions of urban middle classes*). Het Spinhuis/Maklu. Apeldoorn.

Kwaadsteniet, P.I.M., J.F. Jonkhof en S.P. Tjallingii. 2000. *Leve(n)de Stadswateren* (*Living urban waters*). StoWa report 15, Utrecht (STOWA).

Lankamp, P.C. and Valentijn D. ed. 1992. *Monumenten Inventarisatieproject Den Haag 1850–1940* (*Historic building inventory project for The Hague, 1850–1940*). The Hague.

Local Government Economics Research Centre (COELO). 2006. *Atlas van de locale lasten 2006*. Groningen (University of Groningen, Faculty of Economics).

Ministry of Transport, Public Works and Water Management, 2004. *Water in the Netherlands 2004–2005*. The Hague (V&W).

Municipality of Delft. and Delfland Water Board. 2000. *Waterplan Delft: een blauw netwerk* (*Delft Water Plan: a blue network*). Delft (Municipality Delft).

Municipality of Delft, 2004. *Memorandum: afsluitconstructies grachten oostelijke binnenstad* (*water retaining structures for canals in the east of the city centre*. Delft (Municipality Delft).

Municipality of Delft, 2005. *Memorandum: Delft water file*. Delft (Municipality Delft).

Municipality of Delft, Van der Vorm Vastgoed and Delftwonen. 2003. *Masterplan op Hoofdlijnen: Werelds wonen in de Poptahof* (*Master plan outline: Worldly living in Poptahof*). Delft (Municipality Delft).

Municipality of Dordrecht. 2001. *Waterlint door Wielwijk, prijsvraag.*
Waterlandschap van de Toekomst (*Water ribbon through Wielwijk. A competition for the Water Landscape of the Future*). Dordrecht (Municipality Dordrecht).

Municipality of Dordrecht. 2004. *Wielwijk naar het midden en op orde* (*Wielwijk to the centre and in order*). *District outline 2005–2008 + District year programme 2005 Wielwijk*. Dordrecht.

Municipality of Dordrecht, Hollandse Eilanden en Waarden Water Purification Board and De Groote Waard District Water Board. 2002–2003. *Dordrecht Urban Water Plan, Measures Plan 2003–2007, Dordrecht Urban Water Plan, Exploratory Memorandum, Dordrecht Urban Water Plan, Concept Memorandum*. Dordrecht (Municipality Dordrecht).

Municipality of Dordrecht, Stichting Interstede and Woondrecht. 2004. *Dordrecht-West op stoom* (*West Dordrecht gets up steam*, Autumn document 2004). Dordrecht (Municipality Dordrecht).

Municipality of Haarlem. 1999. *Vaststelling Uitvoeringsplan Schalkwijk 2000+* (*Adoption of Schalkwijk Implementation Plan 2000+*). Haarlem (Municipality of Haarlem).

Municipality of Haarlem. 2004a. *Integraal Waterplan Haarlem* (*Comprehensive Haarlem Water Plan*). Haarlem (Municipality of Haarlem).

Municipality of Haarlem, 2004b. Ontwerp Masterplan Water Schalkwijk 2000+ te Haarlem (Draft Schalkwijk Water Master Plan 2000+, Haarlem) main report. Haarlem (Municipality of Haarlem).

Municipality of Haarlem. 2005. *Actualisatie Uitvoeringsplan Schalkwijk 2000+* (*Updated Schalkwijk Implementation Plan 2000+*), *draft*. Haarlem (Municipality of Haarlem).

RIGO.2000. *Wonen in schalkwijk 2015, een wijkperspectief vanuit het wonen* (*Living in Schalkwijk 2015, a view of the district from the residential perspective*). Haarlem (RIGO).

Shaw R. 2002. 'The International Building Exhibition (IBA) Emscher Park, Germany: A Model for Sustainable Restructuring?' In: *European Planning Studies. Volume 10, Number 1, January 1, 2002.*

South Holland South Alliance for Comprehensive Water Management.1999. *Handboek stedelijke waterplannen* (*Urban water plans manual*). The Hague.

Tauw. 2002. *Schetsontwerp watersysteem Schalkwijk (Schalkwijk Water System Outline Design), final.* Deventer.

Tjallingii, S.P. 1996. *Ecological Conditions.* Dissertation TU Delft.

Tjallingii, S.P. 2005. 'Carrying Structures. Urban development guided by water and traffic networks'. In: E. Hulsbergen *et al. Shifting Sense.* Amsterdam (Techne press).

Tjallingii, S.P. and Eijk, P. van. 1999. *Integraal Waterbeheer Delft. Basisstudie voor het Waterplan Delft (Comprehensive water management for Delft. Base study for the Delft Water Plan).* IBN report 421, Executed by BOOM (Delft) and IBN-DLO (Wageningen), Alterra, Wageningen (Alterra).

Tjallingii, S.P. and Ven, F. van de, ed. 2005. *Water in Drievoud (Water in Triplicate).* Eburon, Delft (TU Delft).

Tjallingii, S.P., I.M. van den Top, J.F. Jonkhof & R. Wentink. 2000. *De Blauwe transformatie; innovatie in het stedelijk waterbeheer, een handboek voor de gemeentelijke praktijk (The Blue Transformation; innovation in urban water management, a handbook for municipal practice).* Executed by Alterra (Wageningen) and Tauw (Deventer), Alterra report 092, Wageningen.

Urban Design and Public Housing Department and Rotterdam City Development Corporation. 2005. *Restructuring programme 2005–2015,* Rotterdam (OBR).

Urban Renaissance Agency, 2005. *Profile of UR,* Tokyo (UR).

Urban Renaissance Agency. 2006. *Creation of cities paying special attention to the environment,* Tokyo (UR).

Ven, G.P. van der. 1993. *Leefbaar Laagland, geschiedenis van de waterbeheersing en de landaanwinning in Nederland (Livable Lowland, a history of water management and reclaimed land in the Netherlands).* Utrecht (Uitgeverij Matrijs).

Vries, M.L. de. 1996. *Nederland Waterland (The Netherlands Water Land).* The Hague (Rijksdienst voor de Monumentenzorg).

Zweerink, K. 2005. *Van Pendrecht tot Ommoord (From Pendrecht to Ommoord).* Bussum (THOTH).

INTERVIEWS

Allard Stroo, project leader for District Affairs and Projects, municipality of Delft.

Arthur van Hagen, Municipality of The Hague Urban Management, Sewers and Water Management Department.

Bert Sekeris, Urban District Manager City Development Department, Municipal District Design, municipality of Dordrecht.

Daniël Goedbloed, consultant, water management department, Rotterdam municipal public works department, functional management & policy, Rotterdam.

Dirk Oudshoorn, Spatial Development, Environmental and Economic Affairs Department, Project manager for the city centre, Municipality of Breda.

E. Kelder, Water Official, City Works Sector, municipality of Dordrecht.

Erhard A. Follmi, Strategic Manager (Water), City Management Sector, municipality of Haarlem.

Hanneke Schippers, Programme Manager Urban Development Sector, municipality of Haarlem.

Peter de Vet, project leader for the Museumpark multi-storey car park, Procurement Department, Rotterdam Development Corporation.

Rene van der Werf, project leader for water and soil, District and City Affairs, Environment specialist team.

René van der Werf, project leader for water and soil, District and City Affairs, Environmental professional team.

Rini van Veen, project manager for Civil Constructions / IBU Ingenieursbureau, city works department.

ADDITIONAL READING

Aalbers, C. and Jonkhof, J.F. 2003. *Plannen met principes (Planning on principle)*. Boxtel (Aeneas).

Arda, A., 1999. 'Alternatieve watermodellen Rokin (Alternative water model for Rokin)'. *Gemeenteblad*, dept. 1, no. 311.

Bestuur Regio Utrecht, 2005. *Regionaal StructuurPlan (Regional Structure Plan) 2005–2010*. Utrecht.

Bijkerk, W.O., W.I.J. de Boer, G.A. Marlet, C.M.C.M. van Woerkens. 2003. *Kijk op kantoren 2003 (View on offices 2003)*. Breukelen (NYFER).

BOOM-Duijvestein, 2004. *Waterstructuur Poptahof (Poptahof water structure)*. Delft (BOOM).

BOOM-Duijvestein, P. van Eijk, S. Tjallingii. 2001. *Water in Schalkwijk*. Delft (BOOM).

Bureau of Sewerage Tokyo Metropolitan Government. 2005. *Sewerage in Tokyo 2005*, Tokyo (MLIT).

Centrum Ondergronds Bouwen. 2004. *COB Nieuws 26*. Rotterdam.

Delfland Water Board. 2004. *Beleidsvisie waterplannen, Samen werken aan een duurzame leefomgeving (Policy concept for water plans, working together on a sustainable living environment)*. Delft (Delfland Water Board).

Delfland Water Board. 2004. *Riolering en het Watersysteem (Sewers and the water system), Part 1: Policy memorandum*. Delft (Delfland Water Board).

Delfland Water Board, Municipality of Delft. 2004. *Watersysteemanalyse Delft, samenvatting (Delft water system analysis, summary)*. Delft

Draijer, M., Toorn Vrijthoff, W. van der, H. de Jong. 1998. *Werk aan de Winkel. De toekomst van de winkelmarkt 1995–2015. (Much to do in the shop. The future of the retail market 1995–2015)*. Delft (TU Delft, Faculty of Architecture).

Erasmus, MC. 2005. *Rotterdam wordt beter!, Nieuwbouwproject Erasmus MC (Rotterdam recovers!, New Erasmus MC construction project*. Rotterdam (Erasmus MC).

Ham, W. van der. 2002. 'De Historie (The History)'. In *Water Landschappen, de cultuurhistorie van de toekomst als opgave voor het waterbeheer (Water landscapes, the cultural history of the future as a water management exercise)*. Lelystad (RIZA).

Haruyama, S. 2005. 'Natural disaster and rural planning in Japan: challenge for solution of flooding disasters'. In S. Haruyama, *Urban-rural planning in Asian Mega-Cities; in search of flood control measures in deltaic cities*. Tokyo, pp. 3–13.

Hengeveld, H. and Segeren, W.A. 1984. *Bouwrijp maken van terreinen (Preparing sites for building)*. Deventer (Kluwer).

Infrastructure Development Institute Japan (IDIJ) and Japan River Association, *Rivers in Japan*. Tokyo.

Jinnai, H. 1984. *A spatial anthropology of Tokyo*. Tokyo.

Jinnai, H. 2004. 'Reevaluation of waterfront space in Tokyo – from destruction to regeneration'. In *Regional Design, based on historical and ecological viewpoints*. Tokyo, pp. 50–63

Jun, M. 2006. *How Japanese Basin Management Systems Work, Appropriate Measures for Water-related Infrastructure Making under Rapid Urbanization*. Tokyo.

Jun, M. 2007. *Appropriate Basin Management Systems under Very Rapid Urbanization, strategic approach based on Japan's previous experience*. Tokyo.

Kamiya, H. 2004. 'Regeneration of Tamagawa Canal and water system planning of Tokyo, in: *Regional Design, based on historical and ecological viewpoints,* Tokyo, pp. 64–70.

Können, G.P. 2001. *Climate scenarios for impact studies in the Netherlands.* De Bilt (KNMI).

Lendering, J. 2005. *Polderdenken. De wortels van de Nederlandse overlegcultuur (Polder thinking. The roots of the Dutch consensus culture).* Amsterdam (Athenaeum-Polak & Van Gennep).

Lintsen, H.W. 1980. *Ingenieurs in Nederland in de negentiende eeuw. Een streven naar erkenning en macht (Engineers in the Netherlands in the nineteenth century. A struggle for acknowledgment and power.* The Hague (Martinus Nijhoff).

Louter F. and Savooyen, E. van, 2005. *Parkeren op niveau, De parkeergarage als ontwerpopgave (Parking on a high level, The multi-storey car park as a design issue, Essay: Parkeren, volwaardig onderdeel van integrale gebiedsontwikkeling en exploitatie (Parking, a full part of comprehensive area development and operation).* Bussum.

Marlet, G. and Woerkens, C. van. 2003. *Atlas van Gemeenten 2003 (Municipalities Atlas 2003).* Breukelen (NYFER).

Matsuda, I. 1990. 'Natural Disasters and countermeasures for the Tokyo lowland', in: *Geographical Review of Japan,* Vol. 63 (ser. B), No. 1, 108–119. 1990.

Municipality of Breda. 2001. *De Bredase Waterkrant (The Breda Water News-sheet).* Breda (Municipality Breda).

Municipality of Breda. 2006. *De Nieuwe Mark (The New Mark).* Breda (Municipality Breda).

Municipality of Breda and others. 2001.*Hoofdrapport, Ontwerp Waterplan Breda (Main Report, Draft Breda Water Plan).* Breda (Municipality Breda).

Municipality of Delft, 2005. Delfland Water Board, *Document: Delft water structure concept.* Delft (Municipality Delft).

Municipality of Delft, Van der Vorm Vastgoed and Delftwonen. 2001. *Ontwerp-Masterplan op Hoofdlijnen: Werelds wonen in de Poptahof (Draft Outline Master Plan: Worldly living in Poptahof).* Delft (Municipality Delft).

Municipality of Rotterdam. 2000. *Council meeting on the 2000–2005 Water Plan and 2000–2005 Municipal Sewer Plan.* Council document 2000-685, Rotterdam (Municipality Rotterdam).

Municipality of Rotterdam, Water Board of Delfland, Schieland and Krimpenerwaard and Hollandse Delta. 2000. *Rotterdam Water Plan 2000–2005.* Rotterdam (Municipality Rotterdam).

Municipality of Rotterdam, Hollandse Delta water board and Higher Water Board of Schieland and Krimpenerwaard. 2005. *Rotterdam Water City 2035,* Rotterdam (Episode Publishers).

Municipality of The Hague. 1999. *Water Plan for The Hague 1998–2002, operational section.* The Hague (Municipality of The Hague).

Municipality of The Hague. 1999. *Water Plan for The Hague 1998–2002, policy section.* The Hague (Municipality of The Hague).

Municipality of The Hague. 2002. *Comprehensive urban district description for Segbroek 2002/2003.* The Hague (Municipality of The Hague).

Municipality of The Hague, 2005. *Municipal sewer plan, The Hague 2006–2010.* The The Hague (Municipality of The Hague).

Municipality of Utrecht. 2003. *Masterplan Stationsgebied Utrecht (Utrecht Station Zone Master Plan),* Utrecht (Municipality of Utrecht).

Municipality of Utrecht. 2004. *Milieubeleid voor een leefbare en duurzame stad, Milieubeleidsplan 2003–2008. (Environmental policy for a liveable and sustainable city, Environmental policy plan 2003–2008).* Houten (Municipality of Utrecht).

Municipality of Utrecht. 2004. *Structuurvisie Utrecht (Structural concept for Utrecht) 2015–2030.* Utrecht (Municipality of Utrecht).

Municipality of Utrecht. 2005. *Waterplan Utrecht (Utrecht Water Plan).* Utrecht (Municipality of Utrecht).

Municipality of Utrecht. 2006. *Ontwerp structuurplan Stationsgebied Utrecht (Draft Utrecht Station Zone Structure Plan.* Utrecht (Municipality of Utrecht).

Palmboom & Van den Bout Stedenbouwkundigen. 2004. *Stedenbouwkundig plan Poptahof Delft (Urban development plan for Poptahof, Delft).* Delft.

Project team for the structural concept for The Hague 2020. 2005. *Wéreldstad aan zee, Structuurvisie Den Haag 2020.* (*World city by the sea, Structural concept for The Hague 2020.* The Hague (Spatial Planning Agency).

Project team Perron H2O. 2005. *Perron H2O*, Houten/Wageningen/Utrecht.

Public Space Working Group. 2003 *Pleidooi voor openbare ruimte. De opgave verkend (The case for public space. The task surveyed).* The Hague (Ministry of Housing, Spatial Planning and the Environment).

Rijsberman, M. ed. 2001. *Duurzaam Stedelijk Waterbeheer (Sustainable urban water management).* DuSWat project, Delft Cluster, Delft.

Rotterdam Development Corporation. 2005. *Press release 6 September.* Rotterdam (OBR).

Rotterdam Development Corporation. 2005. *Parkeergarage Museumpark, Rotterdam durft! (Museumpark Multi-storey car park, Rotterdam dares!).* Rotterdam (OBR).

Rotterdam Municipal public works department, 2005. *Schreuder Prize 2005 'Parking underground, space for water'.* Rotterdam (Municipality Rotterdam).

Schot, J.W. 1998. *Techniek in Nederland in de twintigste eeuw, Deel I. Techniek en ontwikkeling, waterstaat, kantoor en informatietechnologie (Technology in the Netherlands in the twentieth century, Part I. Technology and development, water management, office and information technology).* Zwolle (Waanders).

Sorensen, A. 2002. *The making of urban Japan, City planning from Edo to the twenty-first century*, London (Routledge).

Tauw. 2003. *Waterbalansberekeningen Schalwijk te Haarlem (Water balance calculations for Schalwijk, Haarlem), final report.* Deventer.

Urban Design and Public Housing Department and Rotterdam City Development Corporation, 2006. *Profilering Zuidelijke Tuinsteden (Establishing the profile of the southern garden towns).* Rotterdam (OBR).

Van Eijk, P. and Tjallingii S.P. 2003. De Gidsprincipebenadering (The guiding principle approach)'. In P. van Eijk, ed. *Vernieuwen met Water. Een participatieve strategie voor de gebouwde omgeving (Renewing with water. A participative strategy for the built environment).* Dissertation TU Delft, Delft (Eburon).

Vries, J. de and Woude, A. van der, 1997. *The First Modern Economy. Success, Failure, and Perseverance of the Dutch Economy, 1500–1815.* Cambridge (Cambridge University Press).

Water Board De Stichtse Rijnlanden. 2003. *Waterbeheersplan (Water Management Plan) 2003–2007.* Houten.

Witteveen + Bos. 2001. *Eindrapportage, onderzoek naar aanleg van haven en stadsrivier Westflank binnenstad (Final report, study into the construction of a harbour and urban river in the Westflank district of the city centre).* Breda (Witteveen + Bos).

Woud, A. van der. 1987. *Het lege land, de ruimtelijke orde van Nederland 1798–1848.* Amsterdam (Contact).

Index

Color plates

Plate 1 The Water Project 1854 for Rotterdam.

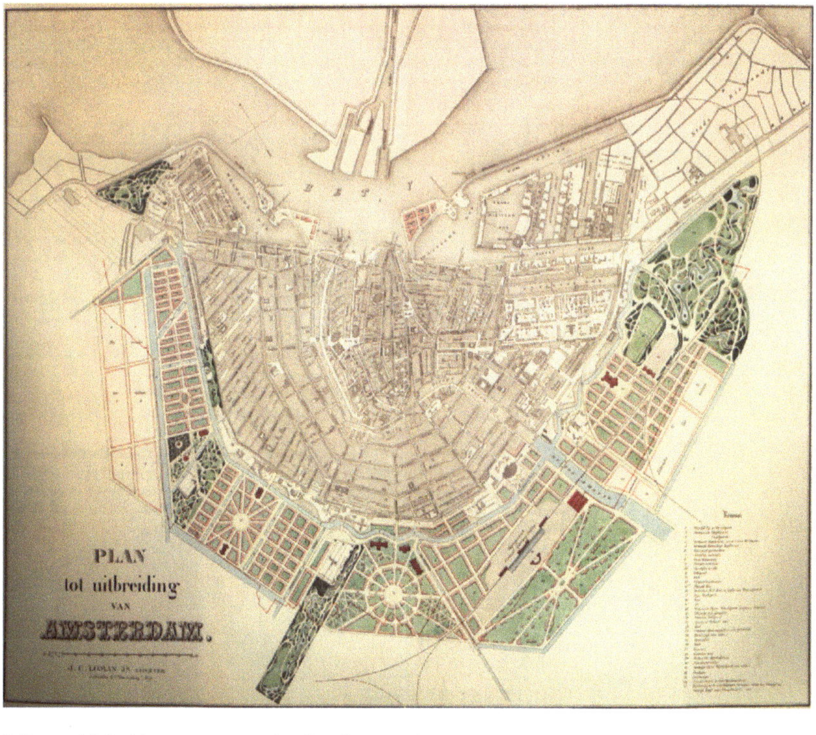

Plate 2 J.G. van Niftrik's expansion plan for Amsterdam.

Plate 3 J. Kalff's expansion plan for Amsterdam.

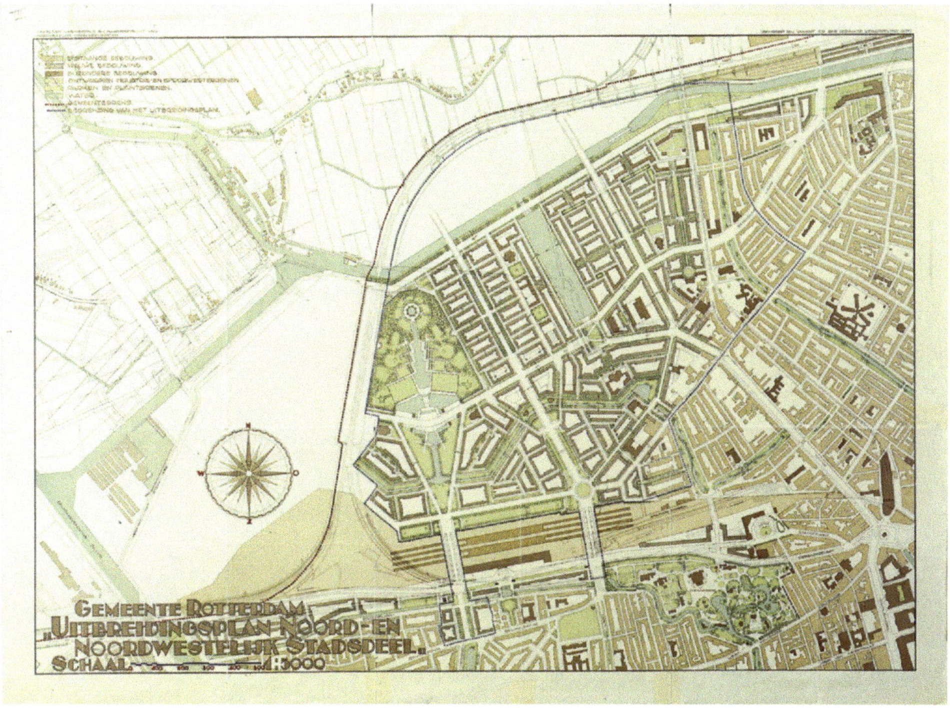

Plate 4 Expansion plan for Blijdorp (1931).

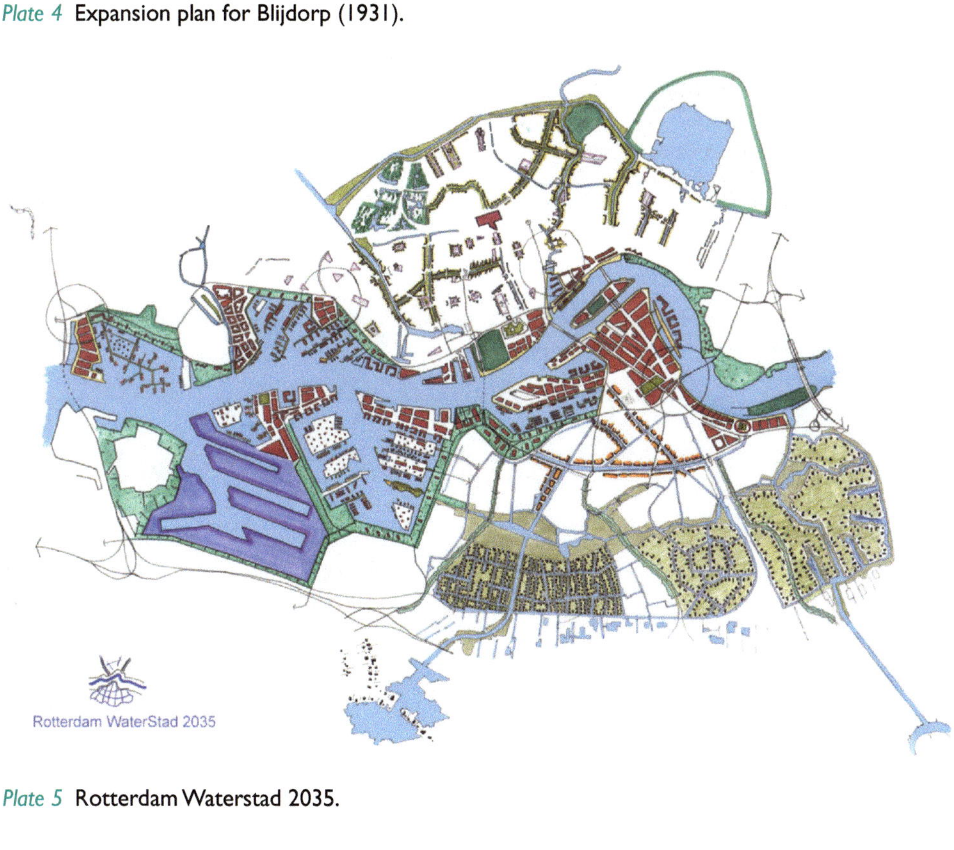

Plate 5 Rotterdam Waterstad 2035.

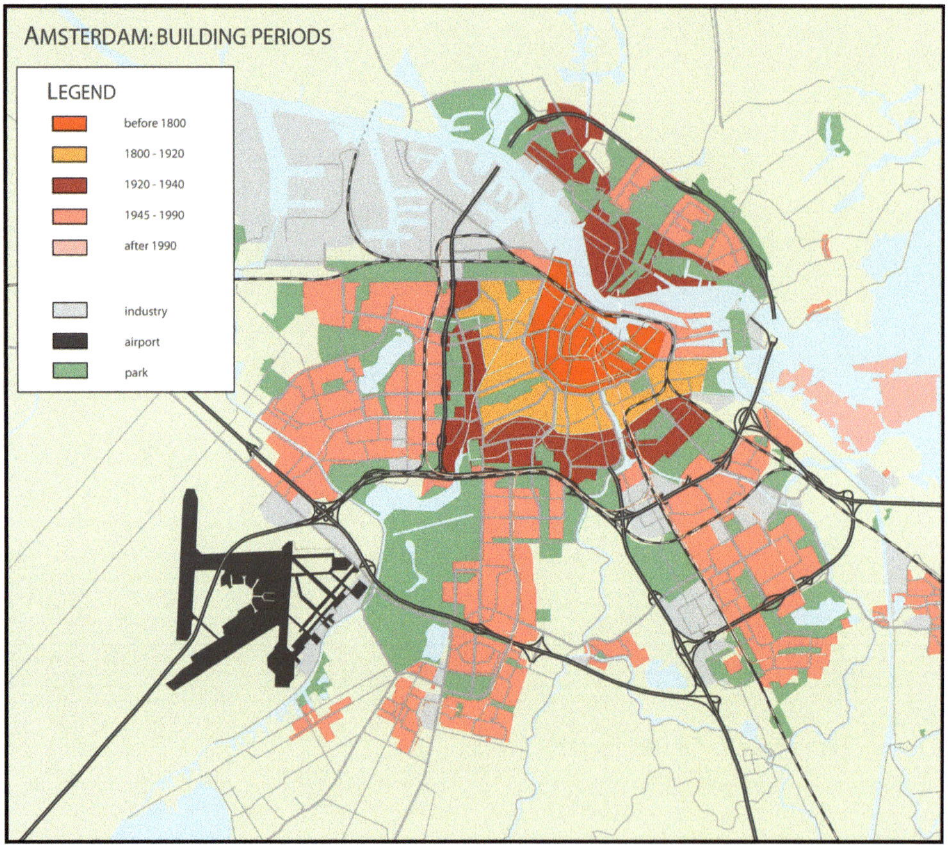

Plate 6 The historic structure of Amsterdam.

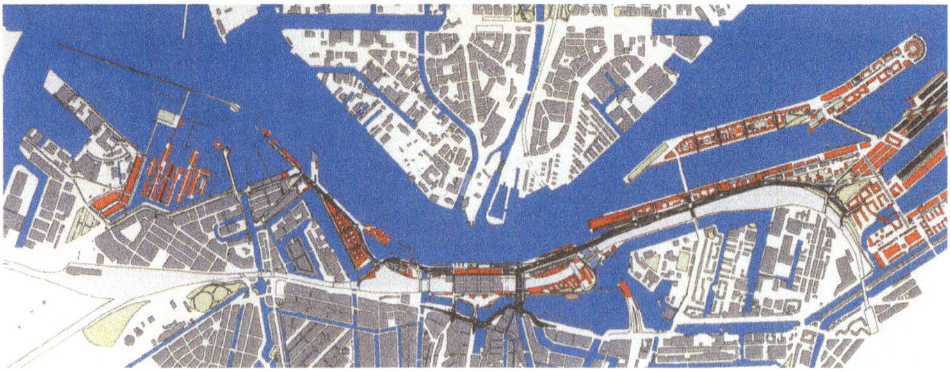

Plate 7 Plan for the IJ bank Amsterdam.

Plate 8 Aerial view of South Rotterdam.

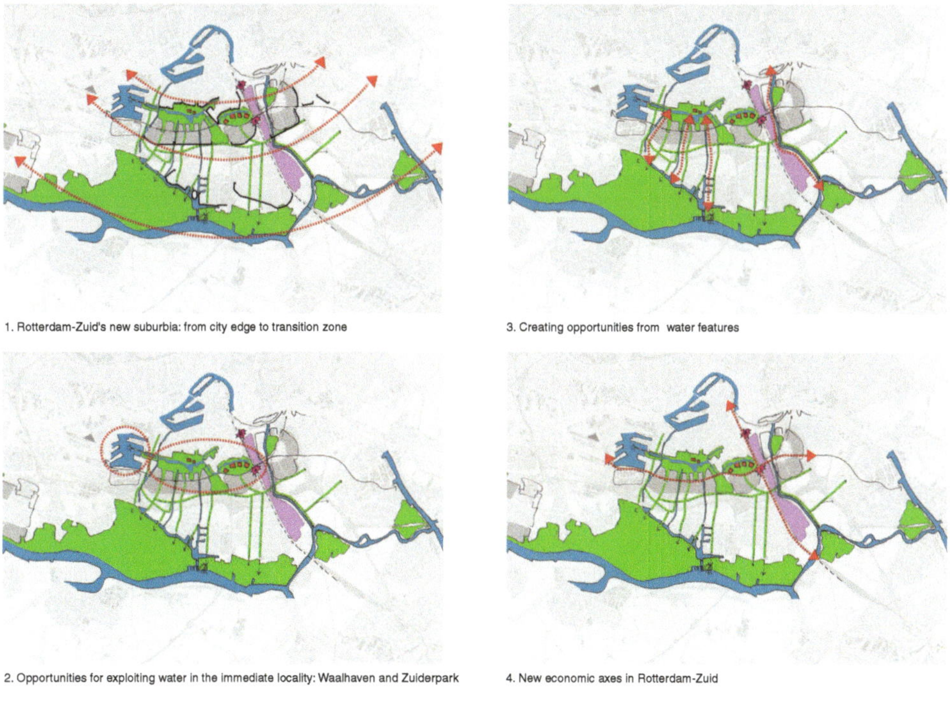

Plate 9 New positions for Rotterdam.

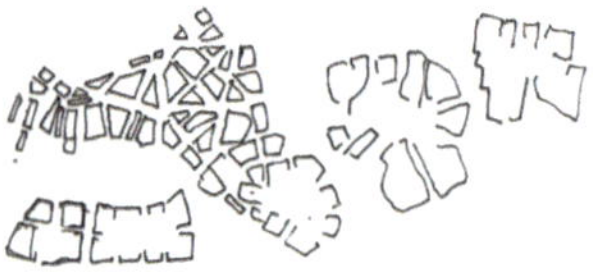

Fragmented city

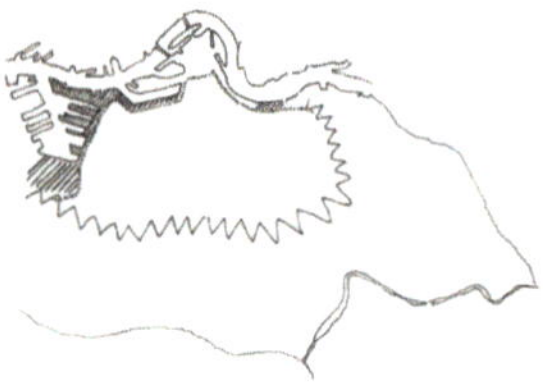

The area of Rotterdam-Zuid behind the harbour, enclosed by infrastructure

Existing water features

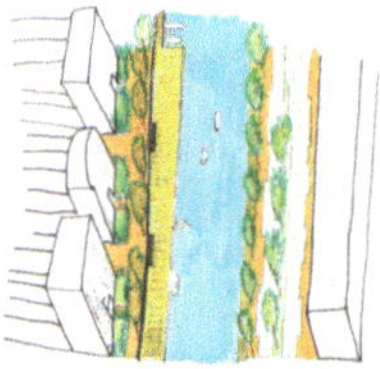

Pleinweg-Strevelsweg

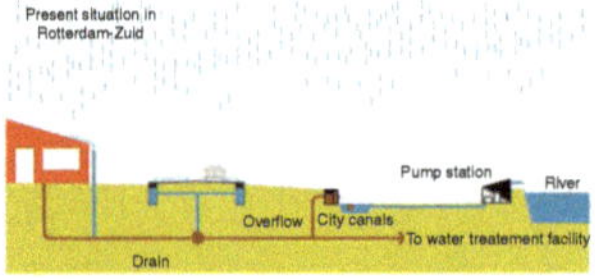

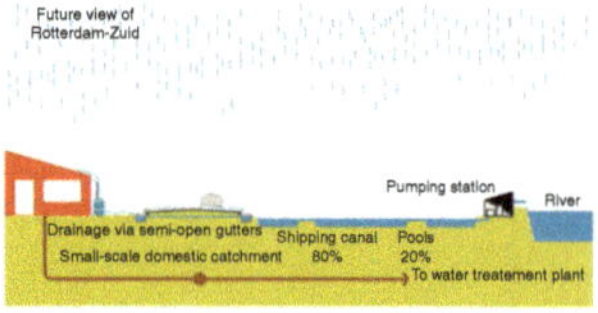

Plate 10 Water map: South Rotterdam canal city.

Plate 11 Ambitions South Rotterdam.

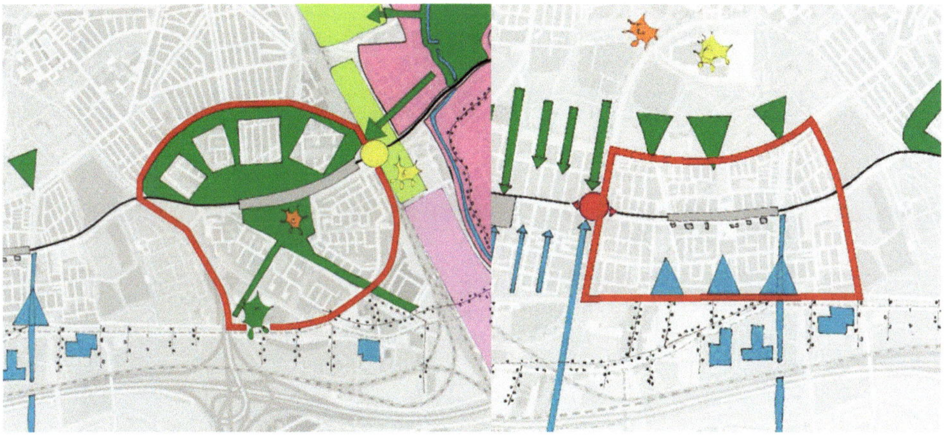

Plate 12, 13 Profiles Lombardijen and Zuidwijk.

Plate 14 The coast of South-Holland.

Plate 15 Aerial photograph of Museumpark (with north downwards).

Plate 16 Flooding of singels.

Plates 17, 18 Artist's impression and cross section of the reservoir.

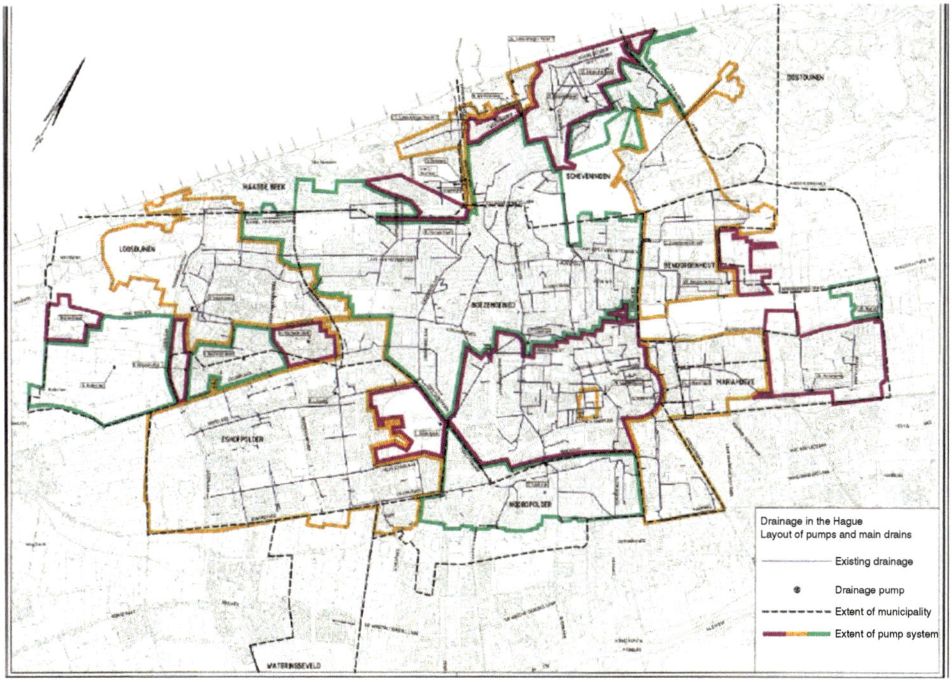

Plate 19 The sewer system in The Hague, pumping areas and mains sewers.

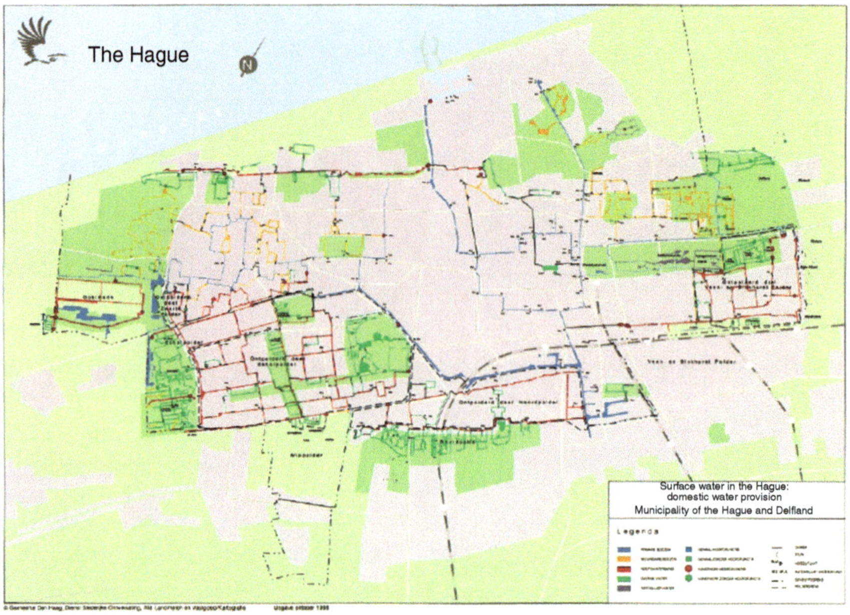

Plate 20 Surface water in The Hague, water management structure.

Plate 21 Planning map of urban design plan with suggested building fields. Poptahof was built in the sixties for 2800 inhabitants in 1019 houses, mostly stacked and 99% rental.

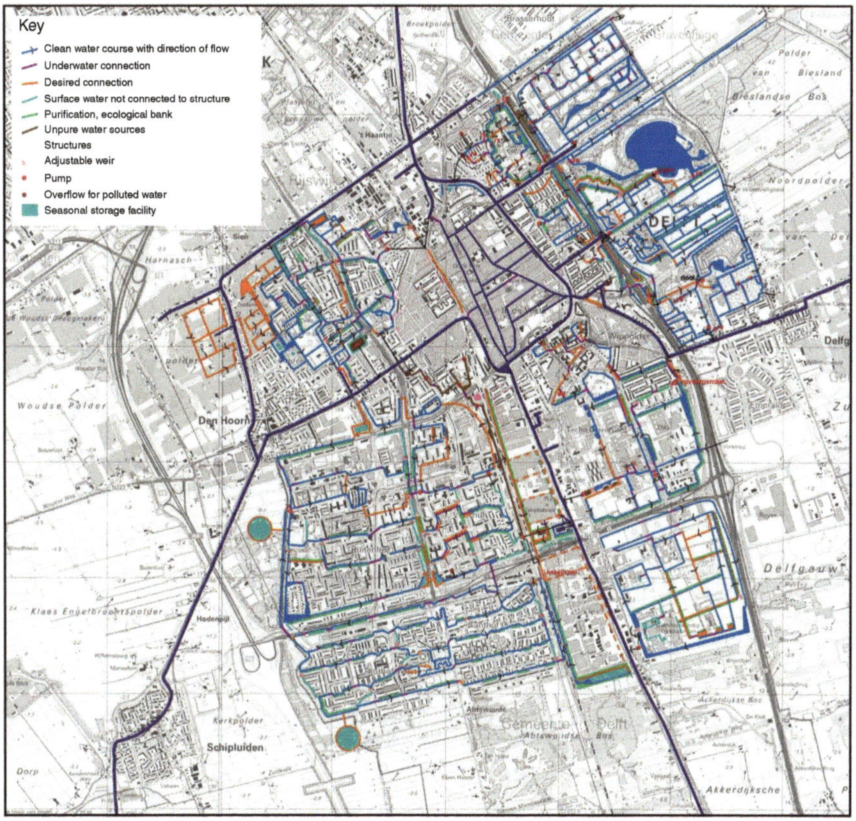

Plate 22 Water structure vision Delft.

Plate 23 Poptahof strategic map.

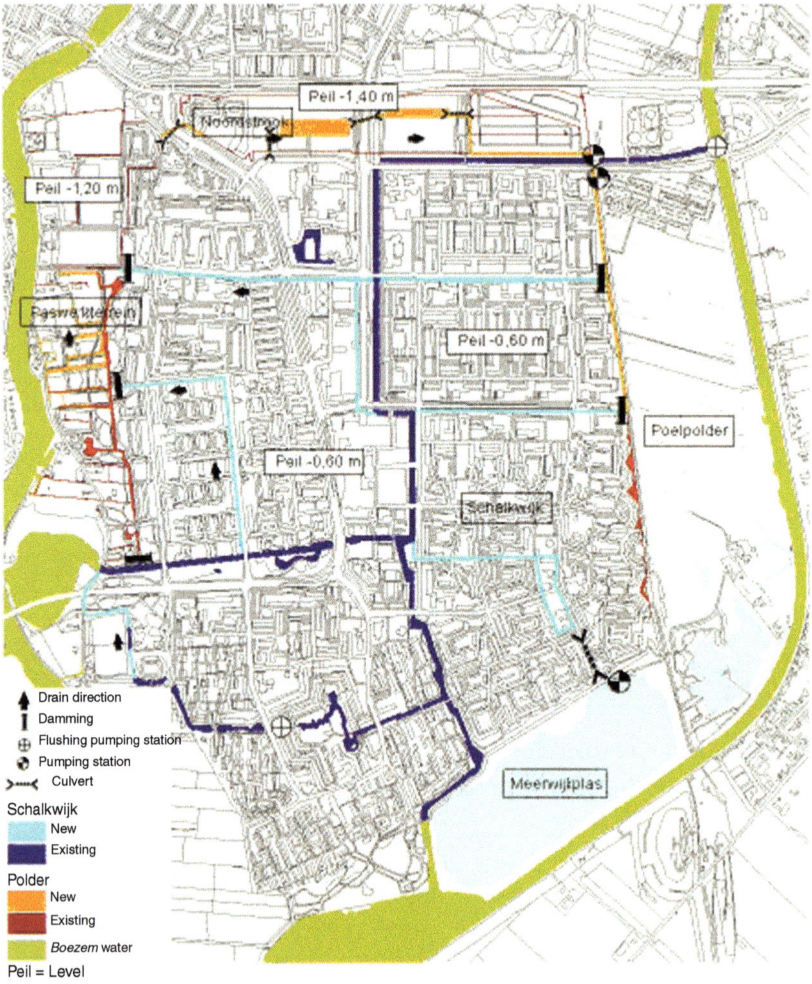

Plate 24 Schalkwijk Base Structure (Light blue is new water; dark blue is existing open water).

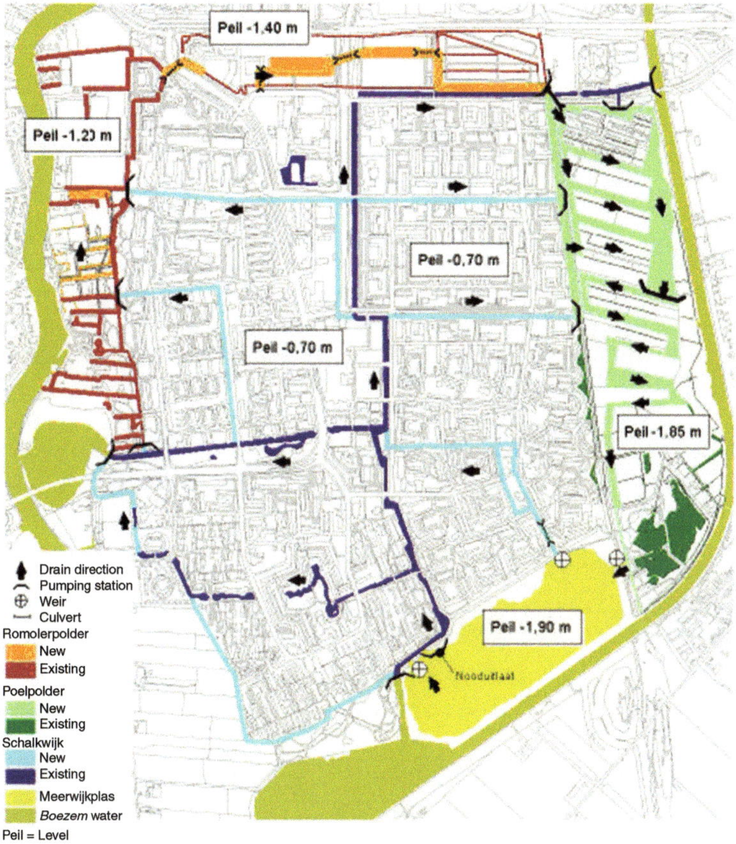

Plate 25 Schalkwijk Outline Design (Light blue is new water; dark blue is existing open water).

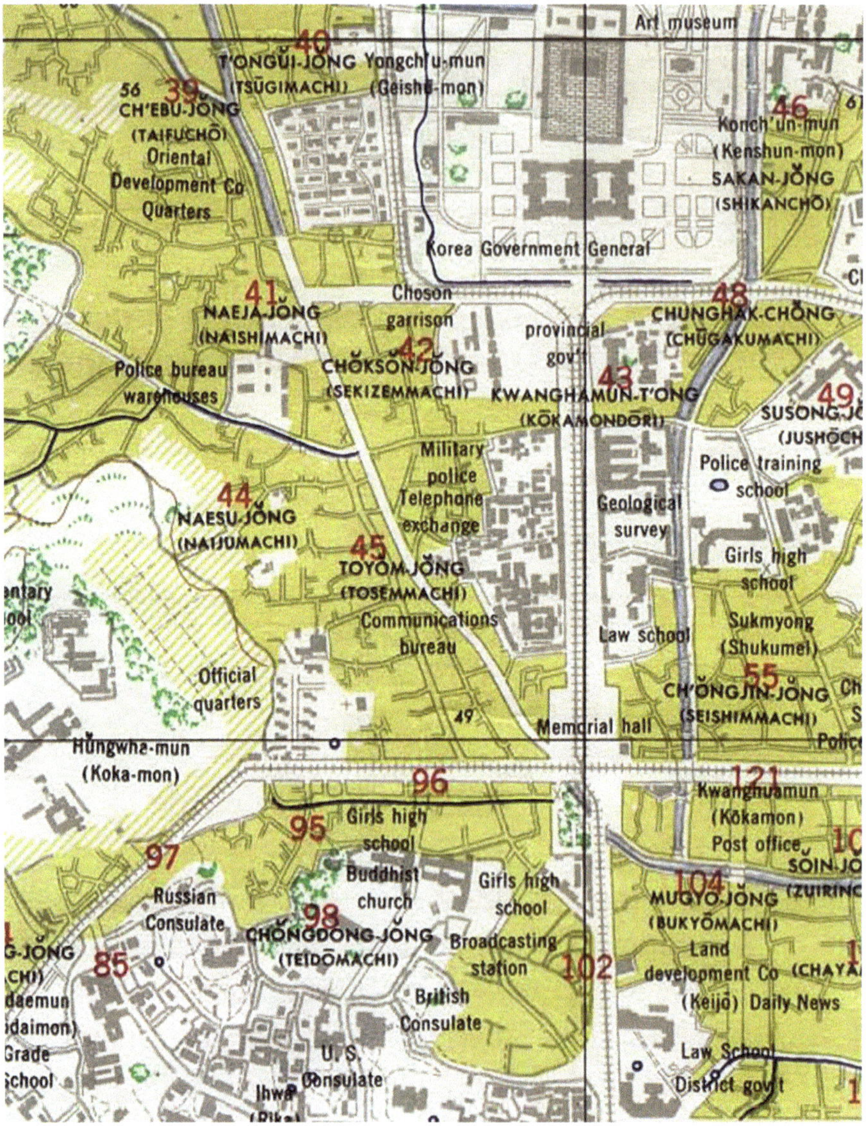

Plate 26 Map existing situation.

Plate 27 Aerial view Japanese polder landscapes.

Plate 28 Nihonbashi Bridge.

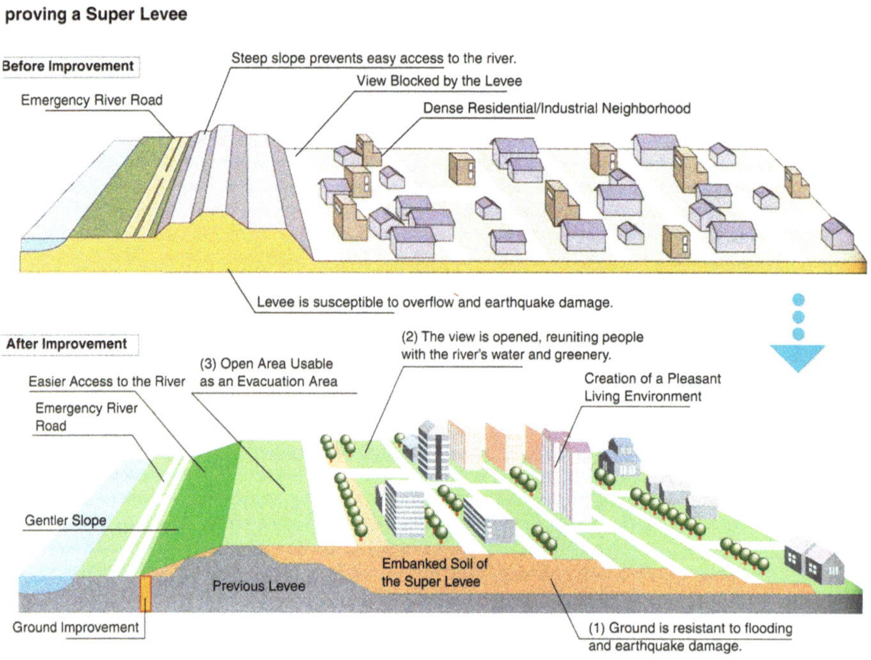

Plate 29 Superlevee.

Plate 30 Masterplan Emscher Park 2010.

Plate 31 Plan River Volume.